METHODS IN MOLECULAR BIOLOGY

For further volumes:
http://www.springer.com/series/7651

For over 35 years, biological scientists have come to rely on the research protocols and methodologies in the critically acclaimed Methods in Molecular Biology series. The series was the first to introduce the step-by-step protocols approach that has become the standard in all biomedical protocol publishing. Each protocol is provided in readily-reproducible step-by step fashion, opening with an introductory overview, a list of the materials and reagents needed to complete the experiment, and followed by a detailed procedure that is supported with a helpful notes section offering tips and tricks of the trade as well as troubleshooting advice. These hallmark features were introduced by series editor Dr. John Walker and constitute the key ingredient in each and every volume of the Methods in Molecular Biology series. Tested and trusted, comprehensive and reliable, all protocols from the series are indexed in PubMed.

Manuscript Instructions https://drive.google.com/file/d/1K4Az5vpAFVzYadJ3jKx2-QSo7Dqa_SNA/view?usp=sharing

Degrons

Methods and Protocols

Edited by

Maria W. Górna

Faculty of Chemistry, University of Warsaw, Warsaw, Poland

Maria M. Klimecka

Faculty of Chemistry, University of Warsaw, Warsaw, Poland

Matylda A. Izert-Nowakowska

Faculty of Chemistry, University of Warsaw, Warsaw, Poland

Humana Press

Editors
Maria W. Górna
Faculty of Chemistry
University of Warsaw
Warsaw, Poland

Maria M. Klimecka
Faculty of Chemistry
University of Warsaw
Warsaw, Poland

Matylda A. Izert-Nowakowska
Faculty of Chemistry
University of Warsaw
Warsaw, Poland

ISSN 1064-3745 ISSN 1940-6029 (electronic)
Methods in Molecular Biology
ISBN 978-1-0716-5507-8 ISBN 978-1-0716-5508-5 (eBook)
https://doi.org/10.1007/978-1-0716-5508-5

This Humana imprint is published by the registered company Springer Science+Business Media, LLC, part of Springer Nature.
The registered company address is: 1 New York Plaza, New York, NY 10004, U.S.A.

Preface

Degrons are specific sequence or structural elements within proteins that mediate their recognition by cellular degradation machinery and are thereby central to controlling protein abundance and cellular homeostasis. This volume aims to provide protocols for the identification, characterization, and application of degrons in a range of organisms from bacteria to humans. In eukaryotes, the most common way to target proteins for degradation is based on recruiting E3 ligases to induce ubiquitination followed by proteosomal degradation. This book starts with reporter assays and tools used in human cells to study the involvement of the ubiquitin-proteasome system or ubiquitin-independent degradation pathways. Next, it presents methods for the chemical and enzymatic synthesis, and characterization of degrons and their mimics. Advanced practical applications of degrons have evolved to aid optogenetic control and live RNA imaging. The dTag system, commonly used for degradation, has been adapted to monitor transcription factors. The book covers methods for other selected eukaryotes, including plants, and describes the use of inducible degradation in nematodes and trypanosomes. The concluding chapters address methods used for studying N- and C-degrons in bacteria, including their emergent application in targeted protein degradation. This collection should be useful to researchers working on protein degradation or using degrons to study or control protein function.

Warsaw, Poland — *Maria W. Górna*
Warsaw, Poland — *Maria M. Klimecka*
Warsaw, Poland — *Matylda A. Izert-Nowakowska*

Contents

Contributors

OMOYEMI AJAYI • *Department of Chemistry, University of Massachusetts Amherst, Amherst, MA, USA*
VYACHESLAV AKIMOV • *Department of Biochemistry and Molecular Biology, University of Southern Denmark, Odense, Denmark*
BUNGO AKIYOSHI • *Centre for Cell Biology, Institute of Cell Biology, School of Biological Sciences, University of Edinburgh, Edinburgh, UK*
PATRICK C. BEARDSLEE • *Department of Chemistry & Biochemistry, University of Delaware, Newark, DE, USA*
BLAGOY BLAGOEV • *Department of Biochemistry and Molecular Biology, University of Southern Denmark, Odense, Denmark*
MARK CARRINGTON • *Department of Biochemistry, University of Cambridge, Cambridge, UK*
HUAXUN FAN • *Department of Biochemistry, University of Illinois at Urbana-Champaign, Urbana, IL, USA*
ETHAN YANG FENG • *Department of Chemistry and Chemical Biology, Harvard University, Cambridge, MA, USA*
JANAINA DE FREITAS NASCIMENTO • *Department of Biochemistry, University of Cambridge, Cambridge, UK*
JULIA M. FRISCHKORN • *Promega Corporation, Madison, WI, USA*
THOMAS M. GEIGER • *Institute of Structural Biology, University of Bonn, Bonn, Germany*
MARIA W. GÓRNA • *Structural Biology Group, Faculty of Chemistry, Biological and Chemical Research Centre, University of Warsaw, Warsaw, Poland*
YU-EN HUANG • *Center for Biophysics and Quantitative Biology, University of Illinois at Urbana-Champaign, Urbana, IL, USA; ReproEngeering T32 Training Grant Program, University of Illinois at Urbana-Champaign, Urbana, IL, USA*
CATHERINE L. INGRAM • *Institute of Structural Biology, University of Bonn, Bonn, Germany*
MIDORI ISHII • *Centre for Cell Biology, Institute of Cell Biology, School of Biological Sciences, University of Edinburgh, Edinburgh, UK*
MATYLDA A. IZERT-NOWAKOWSKA • *Structural Biology Group, Faculty of Chemistry, Biological and Chemical Research Centre, University of Warsaw, Warsaw, Poland*
PRATISTHA KANDEL • *Department of Biological Sciences, University of Delaware, Newark, DE, USA*
ALEXANDRA KLEIN • *Department of Biochemistry, University of Cambridge, Cambridge, UK*
MARIA M. KLIMECKA • *Structural Biology Group, Faculty of Chemistry, Biological and Chemical Research Centre , University of Warsaw, Warsaw, Poland*
STEFAN KNAPP • *Institute of Pharmaceutical Chemistry, Goethe University Frankfurt, Frankfurt am Main, Germany; Structural Genomics Consortium, Buchmann Institute for Molecular Life Sciences, Goethe University Frankfurt, Frankfurt am Main, Germany; German Cancer Consortium (DKTK), German Cancer Research Center (DKFZ), DKTK Site Frankfurt-Mainz, Heidelberg, Germany; Institute of Pharmaceutical Chemistry, Goethe University Frankfurt, Frankfurt, Germany*

TERRENCE CHI-KONG LAU • *Department of Biomedical Sciences, College of Biomedicine, City University of Hong Kong, Kowloon, Hong Kong SAR, China*
CHRISTOPHER LENZ • *Institute of Pharmaceutical Chemistry, Goethe University Frankfurt, Frankfurt, Germany; Structural Genomics Consortium, BMLS, Goethe University Frankfurt, Frankfurt, Germany*
CHUNTONG LI • *School of Pharmaceutical Sciences, Henan Key Laboratory of Nanomedicine for Targeting Diagnosis and Treatment, China Pingyuan Laboratory, State Key Laboratory of Antiviral Drugs, Zhengzhou University, Zhengzhou, Henan, China; Department of Chemistry, New Cornerstone Science Laboratory, Ministry of Education Key Laboratory of Bioorganic Phosphorus Chemistry and Chemical Biology, Tsinghua University, Beijing, China*
JINGHONG LI • *Beijing Life Science Academy, Beijing, China; Center for BioAnalytical Chemistry, Hefei National Laboratory of Physical Science at Microscale, University of Science and Technology of China, Hefei, China; Department of Chemistry, New Cornerstone Science Laboratory, Ministry of Education Key Laboratory of Bioorganic Phosphorus Chemistry and Chemical Biology, Tsinghua University, Beijing, China*
RODRIGO PONTES DE LIMA • *Department of Biochemistry, University of Cambridge, Cambridge, UK*
HANFENG LIN • *The Verna and Marrs McLean Department of Biochemistry and Molecular Pharmacology, Baylor College of Medicine, Houston, TX, USA; Center for NextGen Therapeutics, Baylor College of Medicine, Houston, TX, USA*
ALICE MCDOWELL • *Department of Biochemistry, University of Cambridge, Cambridge, UK*
OSVALDO P. DE MELO NETO • *Aggeu Magalhães Institute, Oswaldo Cruz Foundation, Recife, Pernambuco, Brazil*
PAYEL MONDAL • *Department of Biochemistry, University of Illinois at Urbana-Champaign, Urbana, IL, USA; CureScience Institute, San Diego, CA, USA*
DANIELLE M. N. MOURA • *Department of Biochemistry, University of Cambridge, Cambridge, UK; Aggeu Magalhães Institute, Oswaldo Cruz Foundation, Recife, Pernambuco, Brazil*
QICHANG NIE • *Department of Biomedical Sciences, College of Biomedicine, City University of Hong Kong, Kowloon, Hong Kong SAR, China*
RADOSŁAW P. NOWAK • *Institute of Structural Biology, University of Bonn, Bonn, Germany*
NICOLINE KINGO PEDERSEN • *Department of Biochemistry and Molecular Biology, University of Southern Denmark, Odense, Denmark*
TIEN G. PHAM • *Department of Chemistry, University of Massachusetts Amherst, Amherst, MA, USA*
XIAOLI QI • *The Verna and Marrs McLean Department of Biochemistry and Molecular Pharmacology, Baylor College of Medicine, Houston, TX, USA; Center for NextGen Therapeutics, Baylor College of Medicine, Houston, TX, USA*
CHRISTIAN R. S. REIS • *Department of Biochemistry, University of Cambridge, Cambridge, UK; Aggeu Magalhães Institute, Oswaldo Cruz Foundation, Recife, Pernambuco, Brazil*
KRISTIN RICHING • *Promega Corporation, Madison, WI, USA*
KRISHNA SAXENA • *Institute of Pharmaceutical Chemistry, Goethe University Frankfurt, Frankfurt, Germany; Structural Genomics Consortium, BMLS, Goethe University Frankfurt, Frankfurt, Germany*
KARL R. SCHMITZ • *Department of Biological Sciences, University of Delaware, Newark, DE, USA; Department of Chemistry & Biochemistry, University of Delaware, Newark, DE, USA*

MARTIN P. SCHWALM • *Promega Corporation, Madison, WI, USA*
YOSEF SHAUL • *Department of Molecular Genetics, Weizmann Institute of Science, Rehovot, Israel*
LUYU SHI • *Center for BioAnalytical Chemistry, Hefei National Laboratory of Physical Science at Microscale, University of Science and Technology of China, Hefei, China*
RICHARD STRASSER • *Department of Biotechnology and Food Science, Institute of Plant Biotechnology and Cell Biology, BOKU University, Vienna, Austria*
PATRYCJA E. SZYBOWSKA • *Structural Biology Group, Faculty of Chemistry, Biological and Chemical Research Centre, University of Warsaw, Warsaw, Poland*
ULRIKE VAVRA • *Department of Biotechnology and Food Science, Institute of Plant Biotechnology and Cell Biology, BOKU University, Vienna, Austria*
CHRISTIANE VEIT • *Department of Biotechnology and Food Science, Institute of Plant Biotechnology and Cell Biology, BOKU University, Vienna, Austria*
JIN WANG • *The Verna and Marrs McLean Department of Biochemistry and Molecular Pharmacology, Baylor College of Medicine, Houston, TX, USA; Center for NextGen Therapeutics, Baylor College of Medicine, Houston, TX, USA*
JORDAN D. WARD • *Department of Molecular, Cell, and Developmental Biology, University of California-Santa Cruz, Santa Cruz, CA, USA*
CHRISTINA M. WOO • *Department of Chemistry and Chemical Biology, Harvard University, Cambridge, MA, USA; Associate Member of the Broad Institute, Cambridge, MA, USA*
JIAHUI WU • *Department of Chemistry, University of Massachusetts Amherst, Amherst, MA, USA; Center for Bioactive Delivery in the Institute for Applied Life Sciences, University of Massachusetts Amherst, Amherst, MA, USA; Molecular and Cellular Biology Graduate Program, University of Massachusetts Amherst, Amherst, MA, USA*
WENQING XU • *Department of Chemistry and Chemical Biology, Harvard University, Cambridge, MA, USA*
KAI ZHANG • *Center for Biophysics and Quantitative Biology, University of Illinois at Urbana-Champaign, Urbana, IL, USA; ReproEngeering T32 Training Grant Program, University of Illinois at Urbana-Champaign, Urbana, IL, USA; Department of Biochemistry, University of Illinois at Urbana-Champaign, Urbana, IL, USA; NSF Science and Technology Center for Quantitative Cell Biology, University of Illinois at Urbana-Champaign, Urbana, IL, USA*
YINGYUE ZHANG • *Center for BioAnalytical Chemistry, Hefei National Laboratory of Physical Science at Microscale, University of Science and Technology of China, Hefei, China*
ZHENGUANG ZHAO • *Department of Chemistry and Chemical Biology, Harvard University, Cambridge, MA, USA*

Part I

Reporter Systems and Tools for Studying Degradation in Mammalian Cells

Check for updates

Chapter 1

HiBiT-RR and nLuc^{KO} Tagging Systems for Monitoring Targeted Protein Degradation

Hanfeng Lin, Xiaoli Qi and Jin Wang

Abstract

Targeted protein degradation (TPD) has become a powerful modality in drug discovery. Central to developing TPD molecules like PROTACs is the ability to accurately and sensitively measure the degradation of a target protein. The NanoLuciferase (nLuc) and the complementary split-nLuc system are highly sensitive reporter technologies ideal for this purpose. However, the presence of lysine residues on these tags can lead to ubiquitination of the tag itself, creating potential artifacts where the tag, not the target protein, drives degradation. This chapter details protocols for the generation and use of lysine-free variants, **HiBiT-RR** and **nLuc^{K0}**, where all lysine residues are replaced with arginine. These alternative tools mitigate the risk of tag-driven degradation, thereby increasing the reliability of TPD screening assays. We describe methods for their use in cellular degradation assays with both transiently expressed and endogenously tagged proteins, and validation via immunoblotting.

Keywords Targeted protein degradation (TPD), PROTAC, NanoLuciferase (nLuc), HiBiT, NanoBiT, Lysine, Ubiquitination, High-throughput screening, Reporter assay

1 Introduction

Targeted protein degradation (TPD) is a therapeutic strategy that co-opts the cell's own ubiquitin-proteasome system to eliminate specific proteins of interest (POIs) [1]. The most prominent TPD molecules are proteolysis-targeting chimeras (PROTACs), which are heterobifunctional molecules that simultaneously bind a POI and an E3 ubiquitin ligase, inducing the ubiquitination and subsequent proteasomal degradation of the POI [2].

Developing effective degraders requires robust, sensitive, and high-throughput methods to quantify changes in protein levels. Reporter systems based on NanoLuciferase (nLuc), a small, intensely bright luciferase from deep-sea shrimp, are exceptionally well-suited for this task [3]. A POI can be fused to the full-length nLuc, and its degradation can be monitored by a decrease in luminescence signal. Alternatively, the NanoBiT system uses a split-luciferase

Maria W. Górna et al. (eds.), *Degrons*, Methods in Molecular Biology, vol. 3069,
https://doi.org/10.1007/978-1-0716-5508-5_1,

approach. The POI is tagged with a small 11-amino-acid peptide, HiBiT (High-affinity BiT), which has a sub-nanomolar affinity for the large, complementary LgBiT subunit. When LgBiT is added to cell lysates containing the HiBiT-tagged protein, a bright, active luciferase is formed, allowing for precise quantification of the POI [4]. This system is particularly powerful when HiBiT is knocked into the endogenous locus of a gene using CRISPR/Cas9, allowing for the study of degradation at physiological expression levels [5, 6].

A significant concern with any protein tagging strategy in the context of TPD is the potential for the tag itself to be ubiquitinated. The canonical HiBiT peptide (**VSGWRLFKKIS**, hereafter HiBiT-KK) contains two lysine residues. Similarly, wild-type nLuc ($nLuc^{WT}$) contains seven lysine residues. If a PROTAC brings the tagged POI into proximity with an E3 ligase, these accessible lysines on the tag could be ubiquitinated, potentially leading to degradation that might not occur with the untagged, endogenous protein. This could lead to false positives, where a compound appears to be a potent degrader when it is only degrading the tag [7].

To address this potential artifact, we have developed lysine-free versions of these tags [8]. In **HiBiT-RR**, the two lysines are replaced with arginines (VSGWRLF**RR**IS). In the no-lysine nLuc (**$nLuc^{K0}$**), all lysine codons are mutated to arginine codons. This chapter provides detailed protocols for using these modified tags to reliably monitor protein degradation, minimizing the risk of tag-based artifacts.

2 Materials

2.1 Plasmids and Molecular Cloning

1. Expression vectors for mammalian cells (e.g., pcDNA3.1 with CMV promoter).
2. Plasmids encoding the protein of interest tagged with HiBiT-RR or $nLuc^{K0}$. For example, pcDNA3.1-BTK-HiBiT-RR and pcDNA3.1-$nLuc^{K0}$-RIPK1.
3. High fidelity DNA polymerase for PCR (e.g., KOD One™ PCR Master Mix, Sigma Aldrich, KMM-101NV).
4. Homologous recombination-based seamless cloning kit (e.g., ClonExpress Ultra One Step Cloning Kit V2, Vazyme, C116).
5. Reagents for plasmid amplification: *E. coli* DH5α competent cells, agar plates, LB media, antibiotics, and plasmid MiniPrep kit.

2.2 Cell Lines and Culture

1. For transient overexpression end-point measurement: HEK293T cells (e.g., ATCC, Cat. No. CRL-11268).

2. For endogenous HiBiT-RR knock-in end-point measurement: CRISPR-edited cell lines with endogenous N- or C-terminal tagging of a target protein, e.g., HEK293 HiBiT-RR-BRD4 cells.
3. For kinetic measurement of HiBiT-RR-tagged protein degradation: HEK293 cells stably expressing LgBiT (Promega, N2672) or transiently express LgBiT through transfection.
4. Complete growth medium: Dulbecco's Modified Eagle's Medium (DMEM) supplemented with 10% fetal bovine serum (FBS), 100 U/mL penicillin, and 100 μg/mL streptomycin.
5. Dulbecco's Modified Eagle's Medium (DMEM) without phenol red supplemented with 10% fetal bovine serum (FBS), 100 U/mL penicillin, and 100 μg/mL streptomycin.
6. Transfection reagent (e.g., Lipofectamine 3000, Thermo Fisher Scientific).

2.3 Reagents for Luminescence Assays

1. Nano-Glo HiBiT Lytic Detection System (Promega, Cat. No. N3030), which contains LgBiT Protein, Nano-Glo® Lytic Buffer, and Nano-Glo® Lytic Substrate.
2. Nano-Glo Live Cell Assay System (Promega, Cat. No. N2011), which contains Nano-Glo® Live Cell Substrate and LCS Dilution Buffer.
3. (Optional) Nano-Glo Endurazine Live Cell Substrate (Promega, Cat. No. N2570).
4. Degrader compounds (PROTACs, molecular glues) dissolved in DMSO to a stock concentration of 10 mM.

2.4 Reagents for Immunoblotting

1. RIPA Lysis and Extraction Buffer.
2. Protease and Phosphatase Inhibitor Cocktail.
3. BCA Protein Assay Kit.
4. Primary antibodies specific to the target protein (e.g., anti-RIPK1, anti-BRD4) and a loading control (e.g., anti-β-actin, anti-GAPDH).
5. HRP-conjugated secondary antibodies (e.g., Goat Anti-Rabbit IgG (H+L)-HRP).
6. ECL substrate solution.

2.5 Equipment

1. Luminometer or multimode plate reader with luminescence detection capability (e.g., BioTek Synergy H1).
2. Standard cell culture equipment: incubator (37°C, 5% CO_2), biosafety cabinet, and centrifuges.
3. Solid white, opaque 96-well assay plates with treated surface.
4. Standard equipment for SDS-PAGE and Western blotting.

3 Methods

3.1 Generation of Lysine-Free Constructs

1. Molecular cloning: Use a standard site-directed mutagenesis kit to introduce N- or C-terminus HiBiT-RR or nLuc^{K0} into the target protein open reading frame (ORF).
   For HiBiT-RR, due to the small size of HiBiT-RR, it could be easily introduced as an insertion mutation using a pair of long primers through one round of site-directed mutagenesis. An example of introducing C-terminus HiBiT-RR into the BTK is illustrated in Fig. 1.
   For nLuc^{K0}, it is laborious to systematically mutate all lysine codons in the nLuc sequence to arginine codons. It is recommended to directly order nLuc^{K0} gene through gene synthesis service provider.
2. Transformation: Transform the mutated plasmids into competent *E. coli* strain (e.g., DH5α for normal vectors, or NEBstable for lentiviral vectors) for amplification.
3. Verification: Isolate plasmid DNA from several colonies and confirm the desired mutations and the integrity of the entire open reading frame by Sanger sequencing or whole plasmid sequencing.

3.2 Cellular Degradation Assay (Transient Overexpression)

1. Cell seeding and transfection: The day before transfection, seed HEK293T cells in a 6-well plate so they reach 70%–90% confluency on the day of transfection.

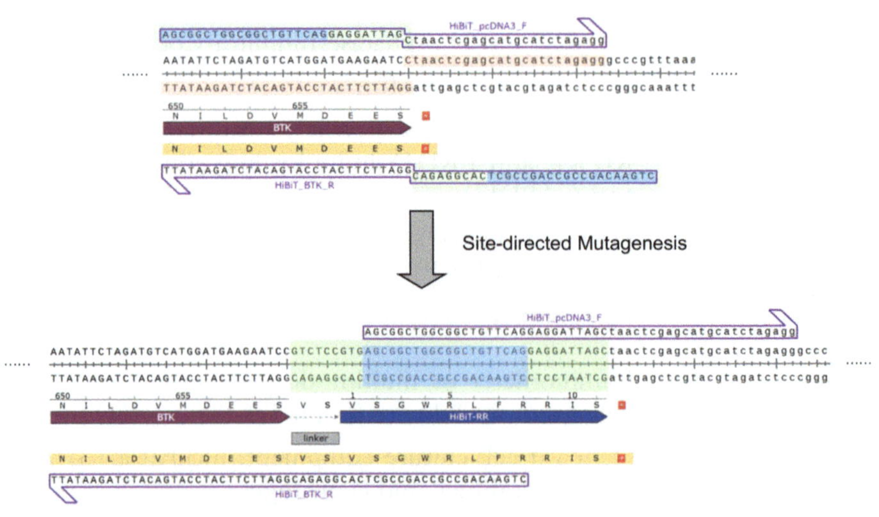

Fig. 1 Introducing HiBiT-RR tag with a short linker (green region) into the target gene ORF via site-directed mutagenesis. The rules-of-thumb of primer design are: (1) Tm of primer annealing regions (highlighted in orange in the upper panel) should be ideally >60°C and (2) region of homologous recombination should be 15–20 bp (highlighted in blue in the lower panel)

2. Transfect the cells with the desired expression plasmid (e.g., pcDNA3.1-BTK-HiBiT-RR or pcDNA3.1-nLuc^{K0}-RIPK1) using a lipid-based transfection reagent (e.g., Lipofectamine 3000) according to the manufacturer's protocol (*see* **Note 1**).
3. Plating for assay: 24 h after transfection, trypsinize and resuspend the cells in fresh complete growth medium without phenol red with 10% FBS. Count the cells and seed them into a white, opaque 96-well plate at a density of 20,000 cells/well in 99 μL of medium (*see* **Note 2**).
4. Compound treatment: Prepare a serial dilution of the degrader compounds in DMSO at 100× the final concentration. Add 1 μL of the 100× compound solution to the appropriate wells. Include several wells of DMSO-only control.
5. Incubation: Incubate the plate for the desired time period (e.g., 24 h) at 37°C, 5% CO_2.
6. Luminescence measurement:
   HiBiT-tagged and nLuc-tagged construct uses different assay buffer (lytic vs live cell, *see* **Note 3**).
   a). For HiBiT-tagged proteins: Prepare 2× Nano-Glo® HiBiT Lytic Reagent by diluting 1:100 (v:v) Nano-Glo lytic substrate and 1:100 (v:v) or 200 nM LgBiT purified protein into the Nano-Glo lytic buffer. Add 100 μL of prepared Nano-Glo® HiBiT Lytic Reagent to each well. Briefly and gently vortex the plate and incubate for 10 min at room temperature to allow HiBiT-LgBiT binding and measure luminescence on a plate reader.
   b). For nLuc-tagged proteins live cell measurement: Prepare 5x Nano-Glo® Live Cell Reagent by diluting the Nano-Glo live cell substrate in the LCS Dilution Buffer (1:19 v:v). Add 25 μL of this reagent to each well. Briefly and gently vortex the plate and measure luminescence on a plate reader (*see* **Note 4**).
7. Data analysis: Normalize the RLU values of compound-treated wells to the average RLU of the DMSO control wells. Plot the normalized percent activity against the log of the compound concentration. Use a nonlinear regression inhibitor vs response three parameters model to calculate the DC_{50} (concentration for 50% degradation) and D_{max} (maximal degradation %) (*see* **Note 5**).

3.3 Cellular Degradation Assay (Endogenous Knock-In)

This protocol is for cell lines where HiBiT-RR has been inserted at the endogenous locus of a target gene (*see* **Note 6**). If kinetic measurement of degradation is required, the cell line must also stably or transiently express LgBiT.

1. Cell seeding: Seed the HiBiT-RR-BRD4 knock-in cells into a white, opaque 96-well plate at 20,000 cells/well in 99 μL of phenol-red free medium. Allow cells to attach overnight.
2. Kinetic measurement (optional): To measure degradation kinetics, add live-cell substrate (1:100 Endurazine v:v) into the CO_2-independent medium and preincubate with cells for 2.5 h in the cell culture incubator.
3. Compound treatment: Add 1 μL of 100× degrader compound DMSO dilutions (e.g., MZ1, dBET6) to the wells.
4. Kinetic measurement continued (optional): Place the plate in the luminometer prewarmed to 37°C and take repeated readings over a time course (e.g., every 10 min for 24 h).
5. Endpoint measurement: For DC_{50} determination, incubate the plate for a fixed period (e.g., 24 h).
6. Lysis and detection: Lyse the cells by adding 100 μL of Nano-Glo® HiBiT Lytic Reagent (Note 3). Briefly and gently vortex the plate and incubate for 10 min and measure luminescence on a plate reader.
7. Data analysis: Analyze the data as described in Sect. 3.2.

3.4 Validation by Immunoblotting (see Note 7)

1. Prepare lysates: Seed cells (e.g., 500,000 cells/well in a 6-well plate) and treat with varying concentrations of the degrader compound for 24 h.
2. Wash cells once with PBS.
3. Lyse cells in 100 μL of ice-cold RIPA buffer containing protease and phosphatase inhibitors. Scrape the cells, transfer to a microfuge tube, and incubate on ice for 30 min.
4. Clarify the lysate by centrifuging at 14,000 × *g* for 15 min at 4°C.
5. Protein quantification: Determine the protein concentration of the supernatant using a BCA assay.
6. SDS-PAGE and transfer: Normalize the protein amounts for all samples, add Laemmli sample buffer, and boil for 5 min. Separate 20–30 μg of protein per lane on an SDS-PAGE gel. Transfer the proteins to a PVDF membrane.
7. Block the membrane with 5% non-fat milk or BSA in TBST for 1 h at room temperature.
8. Incubate with primary antibodies (e.g., anti-RIPK1 at 1:1000 and anti-β-actin at 1:5000) overnight at 4°C.
9. Wash the membrane with TBST three times and incubate with the appropriate HRP-conjugated secondary antibody for 1 h at room temperature.
10. Wash three times again and apply ECL substrate.

11. Visualize the bands using a chemiluminescence imager. The decrease in the target protein band should correlate with the luminescence data.

4 Notes

1. It is extremely important to titrate the amount of plasmid used during the transfection process. Higher transfection amount would significantly reduce or even abolish the degradation potency observed for degrader molecules. Each time we work with a new target protein, we would at least titrate three different levels of plasmid (e.g., 50, 100, 150 ng) to understand the upper limit of transient transfection amount (Fig. 2). Ideally, it is recommended to perform western blotting to confirm the nLuc-tagged protein expression level after transfection should match with the endogenous protein level to avoid false negative result. Optimize the plasmid amount that generates luminescence data correlating well with the Western Blot DC_{50} value before applying to large scale screening.
2. Using medium with phenol red will reduce the luminescence intensity by half. Using transparent 96-well plate will reduce the luminescence intensity to 1/10.
3. Assay format:
   Lytic vs. live-cell: A lytic assay measures the total protein present at the end of an experiment by breaking open the cells. This typically results in homogenous lysate that leads to tighter error bars. A live-cell assay uses the same substrate to measure protein levels in intact, living cells. We observe several fold higher luminescence intensities in the live cell format, but the intensity decays very fast ($t_{1/2}$ ~10 min) if measuring in the

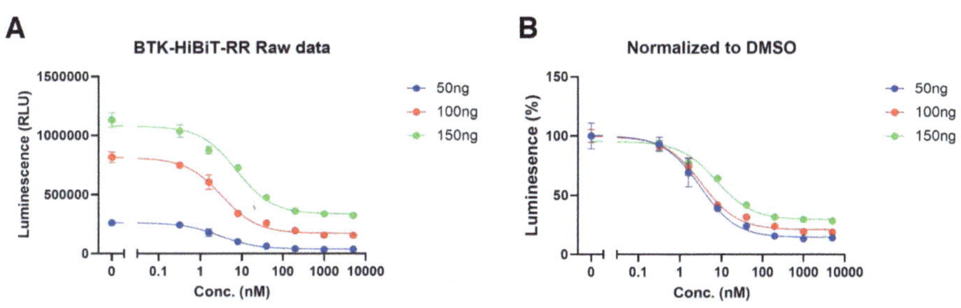

Fig. 2 A representative BTK-HiBiT-RR degradation assay plasmid titration result with unprocessed raw data (**A**) and DMSO-normalized data (**B**). For 50 and 100 ng plasmid curves, both showed similar D_{max} and DC_{50}, but 150 ng plasmid leads to right- and up-shifted curve, indicating the saturation of degradation machinery and therefore underestimation of both DC_{50} and D_{max}

complete media. Using LCS dilution buffer will slow down the decay process ($t_{1/2}$ ~30 min).

HiBiT-RR-tagged experiment typically uses lytic format, unless LgBiT is coexpressed (stably or transiently) in the cell.

Endpoint vs. kinetic: An endpoint assay is a single measurement taken at a final time point (e.g., 24 h) to determine the final outcome (DC_{50}, D_{max}). A kinetic assay involves taking multiple measurements over time to observe the rate of degradation. Kinetic assays require a live-cell format.

4. The $nLuc^{K0}$ protein has significantly lower intrinsic enzymatic activity compared to $nLuc^{WT}$. When using $nLuc^{K0}$ constructs in cellular assays, the luminescence signal will be weaker. This can typically be compensated for by increasing the gain (PMT setting) on the luminometer. Although the protein may be less stable in vitro, it is sufficiently stable in the cellular environment for reliable degradation measurements.
5. Compound-dependent effects: The contribution of the tag to degradation can be compound-dependent. For some PROTACs, the degradation profiles for HiBiT-KK and RR tags may be identical. For others, a difference in kinetics or potency may be observed. Comparing the two systems can provide mechanistic insight into how a specific PROTAC orients the target protein within the ternary complex.
6. Overexpression vs. endogenous systems: Transient overexpression can sometimes lead to artifacts not seen at physiological expression levels, especially when severely overexpressed. Whenever possible, using a CRISPR knock-in system is preferred, as it more accurately reflects the biology of the endogenous target protein.
7. Orthogonal validation: Reporter assays are excellent for high-throughput screening, but it is crucial to validate key findings with an orthogonal method like Western blotting, which directly visualizes the protein of interest.

References

1. Sakamoto KM, Kim KB, Kumagai A et al. (2001) Protacs: chimeric molecules that target proteins to the Skp1-Cullin-F box complex for ubiquitination and degradation. Proc Natl Acad Sci U S A 98:8554–8559. https://doi.org/10.1073/pnas.141230798
2. Mullard A (2021) Targeted protein degraders crowd into the clinic. Nat Rev Drug Discov 20:247–250. https://doi.org/10.1038/d41573-021-00052-4
3. Hall MP, Unch J, Binkowski BF et al. (2012) Engineered luciferase reporter from a deep sea shrimp utilizing a novel imidazopyrazinone substrate. ACS Chem Biol 7:1848–1857. https://doi.org/10.1021/cb3002478
4. Dixon AS, Schwinn MK, Hall MP et al. (2016) NanoLuc complementation reporter optimized for accurate measurement of protein interactions in cells. ACS Chem Biol 11:400–408. https://doi.org/10.1021/acschembio.5b00753
5. Schwinn MK, Machleidt T, Zimmerman K et al. (2018) CRISPR-mediated tagging of endogenous proteins with a luminescent peptide. ACS Chem Biol 13:467–474. https://doi.org/10.1021/acschembio.7b00549

6. Riching KM, Mahan SD, Urh M, Daniels DL (2020) High-throughput cellular profiling of targeted protein degradation compounds using HiBiT CRISPR cell lines. J Vis Exp (165):e61787. https://doi.org/10.3791/61787
7. Zeng M, Xiong Y, Safaee N et al. (2020) Exploring targeted degradation strategy for oncogenic KRASG12C. Cell Chem Biol 27:19–31. https://doi.org/10.1016/j.chembiol.2019.12.006
8. Lin H, Riching K, Lai MP et al. (2024) Lysineless HiBiT and NanoLuc tagging systems as alternative tools for monitoring targeted protein degradation. ACS Med Chem Lett 15:1367–1375. https://doi.org/10.1021/acsmedchemlett.4c00271

Chapter 2

Identification of a Ubiquitin-Independent Degron by a Reporter Assay

Yosef Shaul

Abstract

Ubiquitin-independent degradation (UID) by the 20S proteasome represents a key pathway for mainly regulating intrinsically disordered proteins (IDPs), yet the degron motifs that govern this process remain poorly defined. In this chapter, we describe experimental strategies to identify UID degrons that mediate degradation by the catalytic 20S proteasome particle. Combining high-throughput 20S substrate assays in vitro with analyses of PSMA3 C-terminal binding proteins, we identified numerous IDPs, including p21, as potential substrates. Using bimolecular fluorescence complementation (BiFC) and split-luciferase reporter assays, we mapped the degron within p21. Systematic mutagenesis and peptide- and fragment-based assays delineated the minimal degron sequence, while CRISPR-Cas9–mediated editing of the endogenous protein confirmed its physiological role in proteasomal turnover. This integrated workflow—spanning degron mapping to functional validation—provides a generalizable framework for uncovering ubiquitin-independent degrons.

Keywords Degron, Ubiquitin-independent degradation, 20S proteasome, p21, PSMA3 trapper, IDP, Reporter assay

1 Introduction

Proteasomal degradation is a fundamental cellular process safeguarding protein homeostasis by selectively dismantling proteins in a timely and controlled manner. At the center of this system lies the proteasome, a large multisubunit protease complex that recognizes proteins marked for destruction and degrades them into short peptides, which are subsequently recycled into amino acids. The most prominent pathway, known as the ubiquitin–proteasome system (UPS), relies on the covalent attachment of ubiquitin chains to substrates, thereby marking them for degradation. This system provides a rapid and highly specific means of eliminating misfolded, damaged, aggregated, or short-lived proteins, preventing their accumulation and potential cytotoxicity. Beyond protein quality control, the UPS is indispensable for diverse biological processes, including cell cycle

Maria W. Górna et al. (eds.), *Degrons*, Methods in Molecular Biology, vol. 3069,
https://doi.org/10.1007/978-1-0716-5508-5_2,

regulation, transcriptional control, signal transduction, and stress responses [1, 2].

In the canonical UPS pathway, degradation is initiated through degron-mediated recognition of substrates by E3 ubiquitin ligases, which catalyze the conjugation of ubiquitin via E2 enzymes. Substrates decorated with polyubiquitin chains are then targeted to the 26S proteasome, a complex composed of the 20S catalytic core capped by the 19S regulatory particle. The 19S subcomplex plays two essential roles: recognizing polyubiquitinated proteins and using ATP-dependent unfoldase activity to open structured substrates, which are then threaded into the 20S catalytic chamber for degradation. This ensures that even compact, globular proteins can be efficiently dismantled in a regulated manner.

While ubiquitination is the predominant signal for proteasomal degradation, it is now clear that not all substrates require this modification. A subset of proteins undergoes ubiquitin-independent proteasomal degradation (UID) at the 26S proteasome. The first substrate identified in this pathway was ornithine decarboxylase (ODC), a key polyamine biosynthesis enzyme recruited to the 26S proteasome by its regulator, antizyme, without ubiquitination. More recently, Midnolin, a nuclear protein that stably associates with the 26S proteasome, has been shown to act as a recruiter for additional UID substrates, further highlighting that the 26S proteasome possesses mechanisms beyond ubiquitin recognition to capture and degrade specific proteins. These examples illustrate that UID is not restricted to the 20S particle but can also occur at the fully assembled 26S proteasome, expanding its functional versatility [3–6].

In parallel, degradation can also proceed via the 20S proteasome, which operates independently of the 19S regulatory cap. Unlike folded proteins, which require ATP-dependent unfolding by the 19S complex, intrinsically disordered proteins (IDPs) or substrates with long unstructured regions (*see* **Note 1**) are naturally flexible and can directly enter the catalytic chamber of the 20S proteasome. This bypasses ubiquitin tagging and ATP consumption, enabling a streamlined route for eliminating aggregation-prone or oxidatively damaged proteins, particularly under stress conditions where ATP or ubiquitin availability may be limited. Accessory factors such as NQO1 have been proposed to facilitate substrate association with the 20S proteasome, further broadening the spectrum of proteins processed through this pathway [7, 8].

A degron is a specific region or sequence within a protein that acts as a signal for targeted protein degradation [2]. Degrons can be short amino acid sequences, structural motifs, or exposed residues found anywhere within the protein, including N- and C-termini, and their presence determines the protein's susceptibility to degradation by cellular machinery. These signals are crucial for regulating protein turnover, as components of the UPS or other degradation pathways recognize them. Degrons may function ubiquitin-dependent—serving as sites for polyubiquitination that direct proteins to the proteasome—or independently of ubiquitin, guiding proteins toward alternative degradation mechanisms. Degrons play an essential role in maintaining cellular homeostasis and controlling diverse cellular processes by modulating the degradation rate of proteins [2, 9, 10].

Identifying the degrons is expected to yield key insights into proteostasis mechanisms and the molecular basis of disease pathogenesis. Broadly, degrons can be classified into three major categories (Fig. 1):

1. **E3-degrons**: motifs recognized by E3 ligases promote substrate polyubiquitination and subsequent targeting to the proteasome.
2. **26S UID degrons**: substrate motifs mediating protein degradation but lacking intrinsic proteasome-targeting capacity. In these cases, degradation is facilitated through interacting proteins capable of directing substrates to the proteasome (e.g., antizyme-mediated degradation of ODC).
3. **20S UID degrons**: substrate motifs that enable direct interactions with proteasome components.

The latter two categories are characteristic of ubiquitin-independent degradation. While the first two mechanisms are discussed in detail in the review [2], I focus on category III, the 20S proteasomal UID degrons, under this chapter.

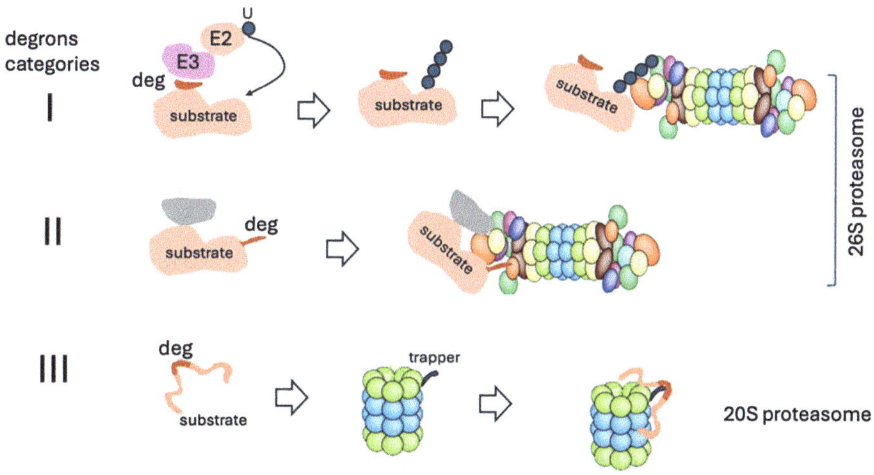

Fig. 1 Categories of degrons and their proteasomal degradation pathways. Schematic representation of three main degron categories and their routes to proteasomal degradation. E3 degrons (I): Substrate degradation is initiated by degron-mediated recognition by E3 ubiquitin ligases, which facilitate ubiquitin (U) conjugation via E2 enzymes. Polyubiquitinated substrates are subsequently recognized and degraded by the 26S proteasome. 26S UID degrons (II): Certain degrons allow direct substrate recognition and engagement by the 26S proteasome without ubiquitination. 20S UID degrons (III): Intrinsically disordered regions (IDRs) containing degrons target substrates to the 20S proteasome in a ubiquitin-independent manner

It is essential to demonstrate that the substrate of interest (SOI) undergoes ubiquitin-independent proteasomal degradation in cells. Although this chapter does not provide detailed protocols, several experimental approaches can be employed. To establish that the proteasome degrades the SOI, degradation should be blocked by treatment with irreversible proteasome inhibitors such as epoxomicin or carfilzomib (more specific than MG132). Conversely, the SOI should not accumulate upon inhibition of lysosomal/autophagic pathways using agents, such as bafilomycin A1 or chloroquine, with or without perturbation of ATG7 or LAMP2A. Once proteasome-dependent degradation has been validated, the next question is whether the process occurs independently of ubiquitination. To test this, cells can be treated with an E1 ubiquitin-activating enzyme inhibitor (e.g., TAK-243/MLN7243), which blocks ubiquitin activation. If inhibition of E1 does not stabilize the SOI, ubiquitin-independent degradation is supported. Alternatively, cell lines bearing a temperature-sensitive E1 enzyme mutant can also be used for validation [11, 12].

There are three significant steps in 20S UID degron identification and biological validation. These include mapping and delineating the 20S proteasome region, recognizing the substrate, termed here the substrate trapper. A few prior substeps validate that the SOI undergoes degradation by the 20S proteasome, as outlined in Fig. 2. Given the space limitation, the methods of these substeps are not covered here but can be found elsewhere [13–15]. In fact, many 20S substrates were already identified using an in vitro system [16, 17]. Interestingly, the identified substrates are mostly IDPs/IDRs, RNA-binding proteins, contain a low complexity region, and prion-like domains [17]. These structural and sequence features should assist in choosing the degradation mechanism of their protein of interest.

Substep 3, as outlined in Fig. 2, is essential in identifying the 20S proteasome subunit mediating the interaction with the substrates in cells. The 20S core particle comprises seven alpha subunits, termed PSMA1-7 and seven beta subunits PSMB1-7 (*see* **Note 2**). The C-terminus of each alpha subunit faces outwards, while the N-terminus faces inwards [18]. We used the BiFC (Bimolecular Fluorescence Complementation) approach (*see* **Note 3**), which is an experimental technique used to study protein–protein interactions in living cells [19]. The method splits a fluorescent protein (like GFP, YFP, or Venus) into two non-fluorescent fragments. Each fragment is fused to a different protein of interest. One protein of interest is fused to the N-terminal half of a fluorescent protein. The other protein is fused to the C-terminal half of the fluorescent protein. If the two proteins interact physically inside the cell, the two fragments of the fluorescent protein are brought close together. This allows the fragments to reconstitute a functional fluorescent protein, emitting fluorescence. The fluorescence signal shows that the two proteins

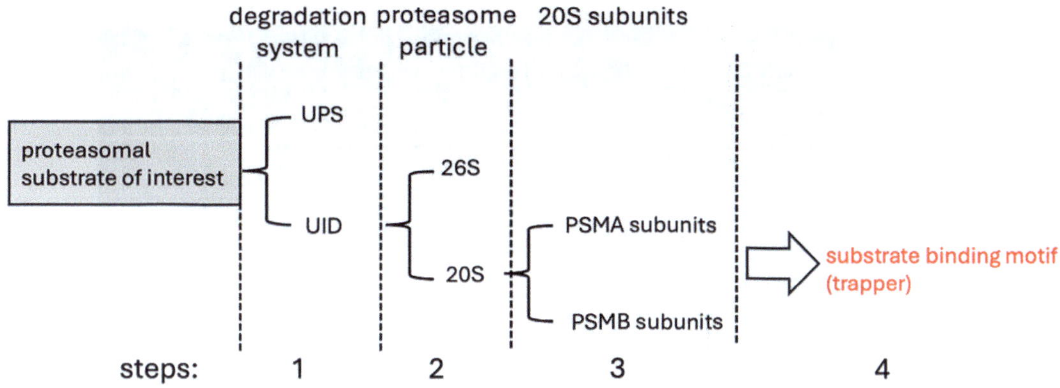

Fig. 2 Stepwise workflow for identifying proteasomal substrate binding motifs. Step 1: A proteasomal substrate of interest enters the degradation system through the ubiquitin–proteasome system (UPS) or ubiquitin-independent degradation (UID). Step 2: The proteasome particle processes the substrate, which may exist as the 26S or 20S core proteasome. Step 3: Within the 20S proteasome, individual PSMA (α-type) and PSMB (β-type) subunits are examined to map potential interaction sites. Step 4: A truncation mutation is used to delineate the subunit region active in substrate binding with the characterization of substrate recognition ("trapper")

interact in a cellular context. Using this approach, we identified many proteins that bind the PSMA3 [17], including p21 (Fig. 3). Theoretically, all the subunits might be considered as the potential target; however, the PSMA3 subunit is the preferential target in several studies [17, 18, 20, 21].

The PSMA subunits differ primarily at their C-termini, residues 186–255 of PSMA3. This region faces outwards in the context of the 20S and 26S proteasomes, and therefore, is accessible to potential substrates. We found that this region is sufficient to bind many putative substrates, most of which are IDP/IDRs [17]. Interestingly, in vitro 20S degradation assay revealed that many of the C-terminus PSMA3 binding proteins are 20S substrates. The substrate "trapper" is essential in identifying the 20S UID degrons.

The second major step is the delineation of the degron sequence. The framework of this step is outlined in Fig. 4. This figure illustrates two complementary strategies for identifying degron motifs—short peptide sequences within a protein that signal for degradation. In Strategy 1 (peptide array-based screening), overlapping peptide fragments of the SOI are synthesized and immobilized as a peptide array. These peptide fragments are incubated with a tagged degron-binding protein ("Tag-Trapper"), and binding is detected by immunoblotting with an anti-tag antibody. Positive binding spots indicate regions that may contain degrons. The positive candidate regions are then mapped back to the SOI, and site-directed mutagenesis, followed by binding assays with split luciferase (N-luc and C-luc) reporters, can refine the degron motif sequence.

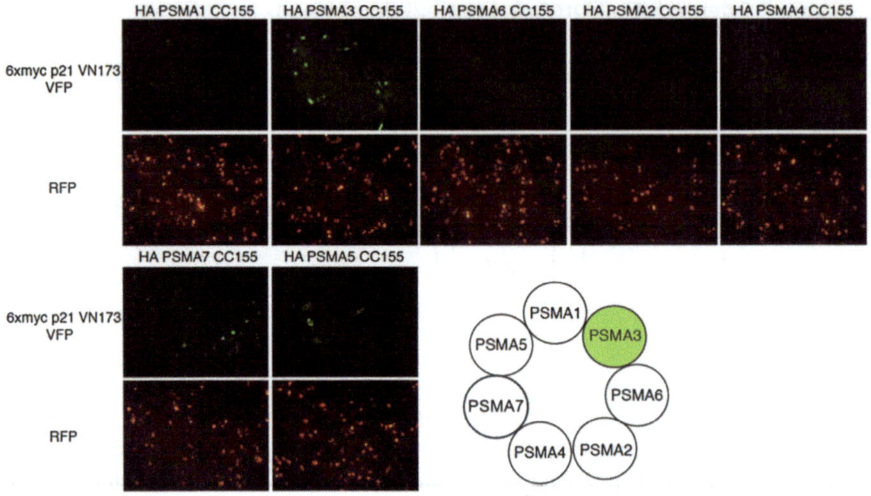

Fig. 3 Interaction of p21 with proteasome α subunits revealed by Bimolecular Fluorescence Complementation (BiFC). Tagged PSMA3 stably expressed in U2OS cells yields a BiFC signal with p21. U2OS cells were infected with lentivirus to generate stable cell lines expressing individual 20S proteasome α subunits. PSMA subunits were tagged with CC155 (C-terminal fragment of Venus fluorescent protein). These cells were transiently transfected with Flag-p21 VN173 (p21 was tagged with the N-terminal fragment of Venus fluorescent protein) and H2B-RFP constructs to label the transfected cells. BiFC signals (green, VFP) were mainly observed for PSMA3, while RFP was a transfection marker. The schematic on the right illustrates the PSMA ring, highlighting the subunits that gave BiFC-positive signals with p21. Images were acquired using a fluorescence microscope with a 10× objective, 48 h post-transfection

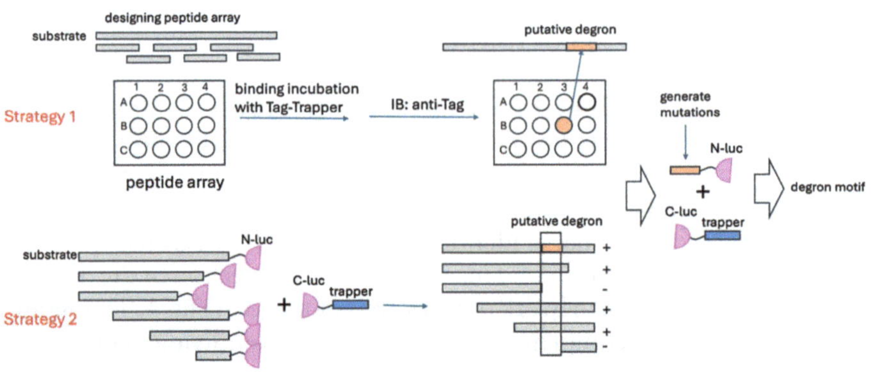

Fig. 4 Strategies for identifying 20S UID degron motifs. Two complementary experimental strategies are shown to identify putative degrons within a substrate protein. Strategy 1 (peptide array-based screening): A peptide array is designed by tiling overlapping fragments of the substrate protein. The array is incubated with a tagged degron-binding protein ("Tag-Trapper") and probed by immunoblotting with an anti-tag antibody. Positive binding spots indicate candidate degron-containing regions, which are then mapped back to the substrate. Mutagenesis of the candidate peptide, followed by binding assays with split luciferase (N-luc and C-luc), identifies the degron motif sequence. Strategy 2 (fragment-based reporter assay): Sequential fragments of the substrate protein are fused to the N-terminal fragment of luciferase (N-luc). Interaction with the degron-binding protein fused to C-luc ("trapper") leads to luciferase complementation and luminescence. Fragments yielding positive signals define regions harboring a degron. Further refinement of overlapping fragments narrows down the degron sequence. Together, these approaches enable systematic identification and validation of degron motifs within substrate proteins

Strategy 2 (fragment-based reporter assay) uses sequential fragments of the SOI fused to the luciferase (N-luc) N-terminal fragment. If the fragment contains a degron, interaction with the degron-binding protein (fused to C-luc, the "trapper") leads to luciferase reconstitution and luminescence. Positive signals identify substrate regions with degron activity, and finer mapping with overlapping fragments narrows the degron sequence further.

These strategies combine screening (peptide arrays) with functional validation (reporter assays). This systematic approach enables precise identification and verification of degron motifs, providing critical insight into how proteins are selectively targeted for degradation within the cell.

The third significant step is physiological validation. A powerful strategy to validate degron function in cells involves using CRISPR-Cas9 genome editing to introduce targeted mutations into the degron sequence. By disrupting this short motif that usually directs proteins for degradation, one can directly test its role in regulating protein stability. Following genome editing, protein synthesis is blocked, and the degradation kinetics of the existing protein pool are measured over time. Protein abundance is then quantified and plotted as $\log_{10}$ (% remaining) versus time to compare stability between wild-type and degron-mutant cells. In control cells, rapid degradation is observed, confirming intact degron activity (Fig. 5).

In contrast, cells harboring degron mutations exhibit delayed degradation, reflected in a longer protein half-life and higher residual protein levels. This shift indicates loss of degron function and stabilization of the protein. Importantly, these edited cells can probe downstream consequences of impaired protein turnover, such as altered signaling, transcriptional regulation, or changes in viability. Thus, combining CRISPR-based mutagenesis with protein stability assays provides a direct and quantitative strategy to validate degron function in a cellular context.

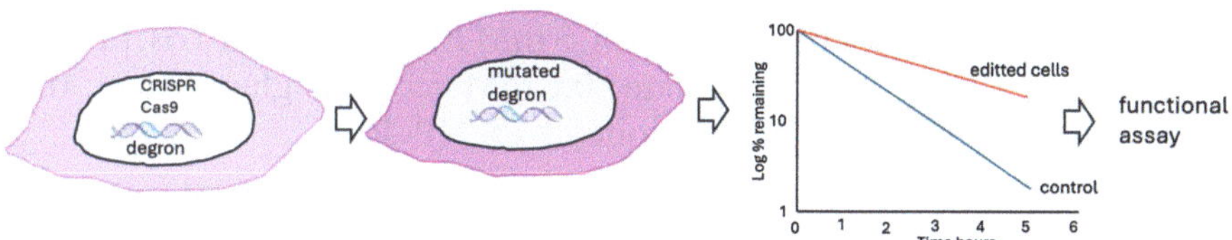

Fig. 5 Functional assay of degron mutation using CRISPR-Cas9. CRISPR-Cas9 genome editing was used to introduce mutations into a degron sequence. In edited cells, the mutated degron impairs protein degradation compared to control cells. After blocking new protein synthesis, protein levels are measured over time and plotted as $\log_{10}$ (% remaining) versus hours. This assay shows protein stability over time, plotted as log % remaining. Control cells (blue line) display rapid protein degradation, while edited cells (red line) exhibit slower degradation, consistent with loss of degron function. Using these edited versus control cells, downstream functional readouts are performed to test the consequence of protein stabilization caused by degron mutation (e.g., viability, signaling, or transcriptional output). Longer $t_{1/2}$ in edited cells predicts increased protein persistence and altered functional outcomes relative to controls

2 Materials

2.1 Tissue Culture and Plasmids

This protocol requires basic tissue culture equipment and supplies, including incubators, laminar flow hoods, centrifuges, inverted fluorescent microscopes, and other standard equipment and plasticware.

1. Plasmids expressing PSMA3 and PSMA5 subunits (Addgene). 6xMyc–p21 (Addgene #87780). N-terminal (Luc1) and C-terminal (Luc2) fragments of split Gaussia luciferase. pBiFC-VN173, pBiFC-CC155 (Addgene #22010, #22015). pcDNA3 vectors with 3×HA tag or Luc1/Luc2 fragments (*see* **Note 4**). Cas9 plasmid (Addgene #42230) for CRISPR editing. Control plasmids containing only tags or luciferase fragments.
2. HEK293, HEK293T, HeLa, and U2OS cell lines growth medium: DMEM (high glucose, 4500 mg/L), 8% fetal bovine serum, 100 u/mL penicillin, 0.1 mg/mL streptomycin.
3. Trypsin solution for passaging the cells: Trypsin (0.25%), EDTA (0.05%) solution B: 2.5 mg/L trypsin, 0.5 mg/L EDTA-Na_2 • $2H_2O$, 8 mg/L NaCl, 0.4 mg/L KCl, 1 mg/L D-glucose, 0.01 mg/L Phenol Red, 0.35 mg/L $NaHCO_3$.
4. **Antibodies**: Anti-HA (BioLegend, or equivalent), anti-Myc (ThermoFisher), anti-p21 (Santa Cruz Biotechnology), anti-p53 Pab1801 (Sigma Aldrich), anti-Flag (Sigma-Aldrich), anti-PSMD1 (Acris), anti-PSMA3 (Thermo Fisher), anti-PSMA4 (Thermo Fisher), anti-β-actin (Sigma-Aldrich), anti-β-tubulin (BioLegend).
5. **Secondary antibodies**: HRP-conjugated goat anti-rabbit, anti-mouse.
6. Tissue culture plates—often six-well dishes.
7. Transfection reagents: Polyethylenimine (PEI), 1 mg/mL.
8. Lysis buffer A (for luciferase and immunoblotting): 150 mM NaBr, 5% glycerol, 5 mM EDTA, 25 mM Tris, pH 8.5, 63.4 μM sodium oxalate, 0.6 mM reduced glutathione, 0.4 mM oxidized glutathione, 0.1% NP-40.
9. Lysis buffer B (for coimmunoprecipitation): 20 mM Tris-HCl, pH 7.5, 320 mM sucrose, 5 mM $MgCl_2$, 1% NP-40.
10. Luciferase assay buffer: 25 mM Tris pH 7.75, 1 mM EDTA, 0.6 mM reduced glutathione, 0.4 mM oxidized glutathione, 75 mM urea.
11. Proteasome inhibitors: MG132, Bortezomib (Calbiochem or equivalent).
12. Cycloheximide (CHX) (*see* **Note 5**).

13. Protease inhibitor cocktail: AEBSF (Aminoethyl benzenesulfonyl fluoride hydrochloride), Aprotinin, Bestatin hydrochloride, E64 [N-(trans-Epoxysuccinyl)-L-leucine 4-guanidinobutylamide], Leupeptin hemisulfate salt, Pepstatin A.
14. Coelenterazine (Nanolight).

2.2 SDS-PAGE and Western Blotting

1. Gel-running and transfer apparatus, such as the Bio-Rad Mini-Protean III system with power supply.
2. PBS: 137 mM NaCl, 2.7 mM KCl, 10 mM phosphate buffer pH 7.4.
3. Protease Inhibitor cocktail, 100×, ApeBio K1007 or equivalent (*see* **Note 5**).
4. SDS-PAGE gels (10%).
5. Prestained protein molecular weight standards (Bio-Rad or equivalent).
6. 3× Laemmli protein gel sample buffer: 6% (w/v) SDS, 30% (v/v) glycerol, 0.1875 M Tris-Cl pH 6.8, 15% (v/v) β-mercaptoethanol, 0.03% (w/v) bromophenol blue.
7. 10× TG (Tris-Glycine): 0.25 M Tris, 1.92 M glycine, pH about 8.3.
8. Gel running buffer: 1× TG, 0.1% (w/v) SDS.
9. Gel transfer buffer: 1× TG, 20% (v/v) methanol.
10. Ponceau Red (Ponceau S) solution: 0.1% (w/v) Ponceau S in 5% (v/v) acetic acid.
11. Nitrocellulose membrane 0.45 μm.
12. 3MM chromatography (blotting) paper.
13. PBST: PBS with 0.1% (v/v) Tween-20.
14. TBST: 10 mM Tris-Cl pH 7.5, 100 mM NaCl, 0.1% (v/v) Tween-20.
15. Non-fat dry milk powder.
16. Primary and secondary antibodies.
17. Reagents for enhanced chemiluminescence, such as the EZ-ECL detection kit (Biological Industries).
18. A digital imaging camera system, such as ImageQuant LAS4000.

2.3 Molecular Biology Equipment and Supplies

1. Thermocycler.
2. TAE (Tris-acetate-EDTA): 40 mM Tris-acetate (pH 8.3), 1 mM EDTA.
3. Suggested suppliers are: BioTri extraction kit (Bio-Lab), qScript cDNA synthesis kit (Quanta Biosciences), PerfeCTa SYBR Green FAST qPCR master mix (Quanta Biosciences).

4. Ethidium bromide (0.625 mg/mL stock solution).
5. Agarose, molecular biology grade.
6. Agarose gel apparatus.
7. Restriction endonucleases (New England Biolabs or equivalent).
8. Purification kits for DNA fragments from agarose gels, such as the QIAquick® Gel Extraction Kit (Qiagen).
9. DNA polymerases for PCR (*see* **Note 6**).
10. Oligonucleotide primers.

3 Methods

The methods below describe the experimental steps after validating that the 20S proteasome degrades the substrate of interest (SOI). As outlined above, there are three major steps; the first is the identification of the proteasome subunit targeted by the substrate. The second describes the experimental strategies to identify the degron motif, and the third deals with in cellulo physiological validation of the degron function. The methods highlighted here are those used by us in discovering the 20S UID p21 degron [22].

3.1 Identification of Proteasome Subunit Interactions by BiFC

As described above, we used the BiFC (Bimolecular Fluorescence Complementation) approach, which detects protein–protein interactions in living cells. The method splits a fluorescent protein (e.g., GFP, YFP, Venus) into two non-fluorescent halves, each fused to a different protein of interest. If the proteins interact, the halves reassemble into a functional fluorescent protein, producing a detectable signal that directly indicates interaction in a cellular context. Direct visualization of each positive cell can be easily quantified. However, this method might introduce a high background because sometimes fragments reassemble spontaneously without real interaction, leading to false positives, and once GFP fragments reassemble, the complex is stable. It is advised to avoid cotransfection and to transfect cells that stably express one of the fragments to decrease the background.

1. Generation of stable cell lines: Seed the cells of interest, in our protocol, the U2OS cells at ~50% confluency, and infect with lentivirus encoding one of the PSMA subunits fused to CC155. Lentivirions are produced in HEK293T cells by cotransfecting a transfer plasmid (with your gene of interest and promoter) with packaging plasmids (providing viral proteins in trans) and an envelope plasmid (defining tropism, e.g., VSV-G). They are designed as self-inactivating for safety and can stably integrate into host genomes. Work must be done in approved BSL-2+ labs, with institutional biosafety approval, since lentivirus

handling carries infectious risks. Replace media after 24 h and maintain stable expression under selection.

2. Transient transfection: To transfect the cell line stably expressing one of the PSMA subunits, dilute the DNA for each transfection using sterile 150 mM NaCl to reach a final volume of 100 μL. Typically, a total of 4 μg of DNA is used for each well of a 6-well plate. To verify transfection efficiency, include 0.05 μg of a plasmid expressing a fluorescent protein, such as H2B-RFP. Use 0.5 μg Flag-p21 VN173 and H2B-RFP plasmids. Use the corresponding empty vector (in our case, pcDNA3) in control wells. A plasmid without a mammalian promoter, such as pBluescript, can be used to bring the total amount of DNA to 4 μg per tube. Prepare a mix of diluted PEI (Polyethylenimine Polyplus Transfection) for all the tubes. Use 16 μL PEI into 100 μL of 150 mM NaCl per tube. Add 100 μL of the diluted PEI solution to 100 μL of the DNA solution (do not reverse this order). Vortex, and let the tube stand at room temperature for 15–30 min. Add the DNA/PEI mix dropwise to the medium of the cells [23]. There is usually no need to change the HEK293 cells' medium with this transfection procedure. The H2B-RFP is used to monitor transfection efficiency.
3. Fluorescence imaging: Wash cells with PBS. Replace with phenol-red-free medium. Acquire images using a fluorescence microscope with Venus and RFP filters. Identify BiFC signal (green, Venus) in RFP-positive cells, indicating interaction between PSMA subunits and p21.
4. It is recommended that the level of expression be monitored and compared, using western blotting. If antibodies are unavailable, the constructs can be tagged for detection.

3.2 Mapping Degron Motifs

Having identified the 20S proteasome subunit that physically interacts with the substrate, we can use it for the next step of mapping the degron. However, to more precisely map the subunit region mediating this interaction, although not critical, we used the subunit truncation mutants to conduct either the BiFC experiment or to use the split-luciferase assay, a technique to study protein–protein interactions in living cells or in vitro. Like BiFC, it relies on splitting a reporter protein—but here, the enzyme luciferase is divided into two inactive fragments. Each fragment is fused to a protein of interest. The luciferase halves are brought together if the proteins interact, reconstituting an active enzyme. When a luciferase substrate (e.g., luciferin or coelenterazine, depending on the luciferase type) is added, the reconstituted enzyme produces light (bioluminescence), which can be measured with a luminometer. Bioluminescence has a low background compared to fluorescence, making detection very sensitive. Light emission occurs almost immediately after substrate addition, suitable for monitoring dynamic interactions. Complementation can dissociate when proteins separate, allowing the study of transient interactions.

Unlike GFP/BiFC, luciferase needs continuous substrate supply, which may limit long-term or spatial studies. However, it provides a quantitative readout but does not localize interactions within the cell as clearly as fluorescence microscopy. In our study, we used the split-luciferase assay and found that the C-terminus region of the PSMA3 subunit binds the substrate and termed it substrate trapper [17].

Having identified the trapper fragment, we used it to map the substrate degron motif. To this end, we used the peptide array screening experimental strategy and also discuss an alternative fragment-based reporter strategy here.

3.2.1 Strategy 1: Peptide Array Screening

This approach enables simultaneous testing of many overlapping peptide fragments in parallel, allowing rapid identification of candidate degron-containing regions. Arrays can be designed to cover entire proteins or focus on specific domains. However, this method may miss structural or conformational effects because peptides are tested outside the full protein context. In addition, in vitro binding does not necessarily reflect actual degron activity in cells. Therefore, candidate regions identified by peptide arrays require follow-up validation through mutagenesis and functional assays.

1. Spot overlapping peptides (15–20 aa) covering the substrate sequence on a nitrocellulose or slide array (*see* **Note 7**).
2. Incubate with tagged degron-binding protein ("Tag-Trapper").
3. Wash to remove unbound protein.
4. Probe with anti-tag primary antibody and HRP/fluorescent secondary antibody.
5. Detect binding by chemiluminescence or fluorescence imaging.
6. Map positive peptide spots back to the substrate sequence.
7. Refine candidate degron motifs by site-directed mutagenesis and validate with luciferase complementation assay.

3.2.2 Strategy 2: Fragment-Based Reporter Assay

As described above, clone and use sequential fragments of the substrate protein fused to the N-terminal fragment of luciferase (N-luc), and PSMA3 full length or the trapper fragment fused to C-luc. These plasmids will be co-trasfected, and the cells will be assayed for luciferase reconstitution and luminescence. By this strategy, the luminescence reflects actual degron activity in the context of protein fragments, not just binding. Also, fragments maintain more of the native protein environment than isolated peptides. However, it requires cloning and testing multiple DNA fragments, which is more time-consuming than peptide arrays. It is possible that larger fragments may mask degron motifs, while too small fragments may lose structural context.

1. Clone sequential fragments of the substrate into the N-luc vector.
2. Clone the trapper, degron-binding protein, into the C-luc vector.
3. Cotransfect mammalian cells with fragment–N-luc and trapper–C-luc constructs.
4. Incubate 24–48 h.
5. Measure luciferase activity using a luminometer.
6. Positive luminescence indicates degron-containing fragments.
7. Refine using overlapping subfragments and mutagenesis.

3.2.3 Luciferase Complementation Assay

1. Transfect HEK293 cells with Luc1-fused POI fragments and Luc2-fused Trapper.
2. Harvest cells after 48 h, lyse in buffer A with protease inhibitors.
3. Split the extract for both the luciferase assay and immunoblotting.
4. Plate cell lysates into 96-well white plates (30 μL/well).
5. Prepare luciferase substrate (20 μM coelenterazine in assay buffer).
6. Mix lysates with coelenterazine substrate (to final concentration of 20 μM) in 96-well plates.
7. Measure luminescence using the Veritas luminometer.

3.2.4 Western Blotting (to Validate Expression)

1. Denature lysates, run SDS–PAGE, and transfer to nitrocellulose.
2. Probe with relevant antibodies.
3. Detect with EZ-ECL and ImageQuant™ LAS 4000.
4. Quantify bands with ImageJ.
5. Analyze the expression and integrity of constructs.

3.3 Functional Validation of the Degron

Having identified a degron motif, its biological activity in the endogenous protein should be assessed using genome editing. CRISPR-Cas9 genome editing is a powerful strategy to introduce targeted mutations into the degron sequence to validate physiological degron function in cells. Basically, mutating the degron motif is predicted to block protein degradation and increase the protein level. Numerous tools and protocols are available [24], and it is advised to review them to choose the most suitable one for editing the substrate under study. The first step is to sequence the targeted genomic region in the cell line of interest—the obtained sequence is used to design the donor template and the guide RNA. Supply the donor repair template to engineer a specific sequence change so that homology-directed repair (HDR) can incorporate the desired sequence. Editing efficiency depends on Cas9 and gRNA expression, gRNA performance, the cell line, and the cell's propensity to use HDR.

In CRISPR-based genome editing, the donor DNA provides the repair template when you want to introduce a specific sequence change (via HDR). The main choices for donors are single-stranded oligodeoxynucleotides (ssODNs) and short (usually 50–200 nt) single-stranded DNA. ssODNs are suitable for introducing minor edits (point mutations, small tags) and are easy to synthesize and deliver. Double-stranded DNA is used for longer homology arms (hundreds to thousands of bp). To this end, the genomic region of interest should be sequenced.

3.3.1 gRNA Preparation

1. Use a web-based tool such as CRISPOR (http://crispor.org).
2. Design a guide RNA that generates a cut close to the site to be edited. If there are several options for guide RNAs, choose the guide with the highest score for cleavage efficiency, which is predicted to cut a minimum of off-target sites.

3.3.2 CRISPR Genome Editing

Only a fraction of cells are successfully edited when performing CRISPR-mediated genome editing. Several selection methods are commonly used to enrich or isolate those cells, depending on the experimental setup, such as antibiotic selection. In this approach, a selectable marker gene (e.g., puromycin, neomycin, hygromycin resistance) is codelivered with CRISPR components. Cells that use the CRISPR/Cas9 machinery (and marker) survive antibiotic treatment. However, resistant cells are not necessarily edited, so further validation is required. Another method is fluorescence-based selection (FACS). In this protocol, a fluorescent protein reporter is coexpressed with Cas9/sgRNA, and FACS sorts cells to enrich for the high-expression cells. Similar to antibiotic selection, the FACS method enriches for transfected cells rather than directly edited cells. A more direct approach is single-cell cloning. Edited populations are diluted and expanded into single-cell clones. Each clone is then genotyped, ensuring identification of the edited clones [24].

We have reported a scarless selection method based on a cell line's temperature sensitivity [25, 26]. This protocol enriches homology-directed repair (HDR)-dependent edits in a gene of interest (GOI) by coupling the edit with repair of a pre-engineered temperature-sensitive (ts) mutation in an essential gene (e.g., TAF1). Cells that restore the wild-type ts allele can survive at a restrictive temperature, providing a scarless selection system. We have engineered HEK293 $TAF1^{ts}$ and HeLa $TAF1^{ts}$, temperature-sensitive cell lines, as reported [25, 26]. However, here, we describe a protocol for editing the wild-type HEK293 genome.

1. Plate 2×10^5 cells per well in a 12- or 24-well plate (0.5 mL DMEM for 24-well, 1 mL for 12-well). The aim is to reach 50%–60% confluency the next day.
2. Incubate at 37°C, 5% CO_2 for 24 h.
3. Prepare DNA or ssODN mixture for each well as follows: 0.05 μg pEGFP, 0.5 μg pX330 (expressing guide RNA and Cs9), 0.4 μg pEFIRES-p (puromycin resistance), 1 μL of 10 μM ssODN (optional: include pBlueScript plasmid to normalize total DNA to 0.5 μg per well, if needed). Dilute DNA into 150 mM NaCl to a 25–50 μL final volume.
4. Add transfection reagent PEI (Polyethylenimine), but other commercial reagents are also helpful. Dilute PEI (1 mg/mL stock, or PEI-Max) in 150 mM NaCl to achieve an N/P ratio 8. Typically, use 3 μL PEI in 50 μL 150 mM NaCl. Add PEI solution directly to the DNA solution, mix by brief vortexing. Incubate 15–30 min at room temperature. Add the entire DNA–PEI mixture dropwise to each well.
5. Assess the transfection efficiency by EGFP or mCherry fluorescence ≥48 h post-transfection.
6. Select for puromycin resistance as follows: replace medium with 5 μg/mL puromycin 30 h post-transfection (*see* **Note 8**).
7. After 24 h, replace with medium without puromycin.
8. Allow cells to recover and replate as necessary.
9. Trypsinize and reseed into 6 cm dishes after initial selection. To isolate clones from single cells, plate cells in 96-well plates at a calculated density of 0.5 cells/well to enrich for single-cell–derived colonies.
10. Monitor growth and re-feed colonies with 50 μL of fresh growth medium as needed.
11. After 10–15 days, identify and pick individual colonies.
12. Expand stepwise (96-well to 24-well, next to 12-well, and finally to 10 cm plates).
13. Split clones for sequencing, freezing (long-term storage), and culture maintenance (ongoing use).

3.4 Genomic PCR and Sequencing

A few issues must be considered when evaluating the sequence validation of the HDR-edited gene. PCR primers should be designed outside the homology arms so the amplicon spans both HDR junctions; avoid binding across the edited bases. Ensure a size compatible with your platform (and readable Sanger traces if using Sanger). Watch for GC-rich/repetitive regions and pseudogenes/segmental duplications that can misprime or mis-map. If you introduced blocking/silent mutations to prevent recutting, include them in your expected sequence model; they help confirm genuine HDR. Finally, include unedited (WT) DNA and, if available, a synthetic positive control containing the intended edit (*see* **Note 9**).

Cell lysis and DNA preparation:

1. Lyse and resuspend 5×10^4–1×10^5 cells in 50 mM NaOH.
2. Incubate at 95°C for 10 min.
3. Neutralize by adding 2 μL of 1 M Tris-Cl (pH 8.0) and 140 μL of nuclease-free water.
4. Prepare PCR reactions (per tube): 25–50 ng of genomic DNA preparation, add 5 pmol of each primer, 2 μL of 5× PCR master mix. Add water to a final volume of 10 μL.
5. Run PCR under optimized cycling conditions for your primers.
6. Run PCR products on a 2% agarose gel.
7. Excise and purify the correct bands.
8. Submit purified products for Sanger sequencing.
9. Analyze sequence traces using ICE (Inference of CRISPR Edits, Synthego) or equivalent software. Confirm that the intended substitution, insertion, or deletion is present and that both HDR junctions are precise, with no extra bases or microindels. Determine mono- vs biallelic editing and whether the sample is clonal. Mixed chromatogram peaks (Sanger) or allele fractions (NGS) indicate mosaicism or mixed clones. Look for on-target indels on the non-HDR allele(s) and partial HDR outcomes (one junction is precise, the other is not). Screen for large deletions, inversions, or rearrangements near the cut site; consider long-range PCR or long-read sequencing if suspicious.

3.5 Functional Validation

Protein stability should be determined to validate the degron function.

1. Conduct Cycloheximide (CHX) chase assay to measure protein half-life. The concentration of CHX required to block protein translation can vary by cell type. Typically, cells are treated with 500 μM CHX, but this can be too high for some cells; test a range (50–200–500 μM) to balance inhibition versus toxicity.
2. Select sufficient time points to capture decay kinetics (e.g., 0, 0.5, 1, 2, 4, and 8 h, depending on expected half-life).
3. Normalize Western blots to total-protein stains (e.g., Ponceau, SYPRO) or stable housekeeping proteins.
4. Fit log-linear decay curves to estimate half-life (t½), and compare against unedited control cells.
5. The half-life should also be determined in cells treated with CHX and an E1 inhibitor to confirm further that the UID pathway mediates decay. Cells carrying degron-disrupting edits should exhibit prolonged half-life, thereby validating degron function.

4 Notes

1. Methods of IDP definition include, among many experimental approaches, the nuclear magnetic resonance (NMR) spectroscopy. Computational/Bioinformatics approaches predict disorder directly from sequence. Tools like IUPred, PONDR, DISOPRED, and PrDOS analyze amino acid composition, hydrophobicity, charge, and complexity. IDPs are enriched in polar/charged residues and depleted in hydrophobic residues. Additionally, IDP can be biochemically determined [13, 14].
2. The 20S proteasome is the catalytic core particle of the proteasome, and its subunit terminology can be confusing because it comes from two traditions: biochemical naming (α and β subunits) and gene-based nomenclature (PSMA/PSMB numbers).
   α1 = PSMA6
   α2 = PSMA2
   β1 = PSMB6 (catalytic)
   β2 = PSMB7 (catalytic)
   β5 = PSMB5 (catalytic)
   β1i = PSMB9 (immuno form)
3. The advantage of the BiFC method is that it detects direct protein–protein interactions in living cells. Provides spatial and temporal information, namely, where and when interactions happen. Straightforward readout using fluorescence microscopy. Limitations are the reconstitution of the fluorescent protein is usually irreversible, meaning transient interactions may appear more stable than they are. Fusion tags can sometimes affect protein folding or function.
4. To construct the plasmids, amplify DNA fragments of the gene of interest by PCR using primers containing appropriate restriction sites. Separate products on 1% agarose gel, excise bands, and purify using the QIAquick® Gel Extraction Kit. Digest PCR products with restriction enzymes. For fragments <120 bp, order complementary oligonucleotides with cohesive ends (such as Sigma-Aldrich). Ligate DNA fragments into a plasmid of interest, such as pcDNA3 vectors containing 3×HA or luciferase tags. Transform ligation mixtures into competent *E. coli*, purify plasmids using the Qiagen Miniprep Kit, and verify constructs by sequencing. Alternatively, one can use the restriction-free (RF) strategy for cloning [27].
5. CHX treatment inhibits the cellular translation machinery.
6. We used several different polymerases for PCR. We used high-fidelity polymerase such as 2× iPfu (iNtRON Biotechnology) to amplify genomic DNA for sequencing. Terra polymerase (Clontech) is suitable for direct PCR from cells or partially purified DNA. For screening colonies, we used Taq polymerase mixes, such as Larva Red Load Taq Master mix (5×).

7. We used commercially available, preprinted peptide arrays in our work. Many vendors offer tiled libraries of 15–20 amino acid peptides (with five to ten residue offsets), preimmobilized on nitrocellulose or functionalized slides. This option is the fastest, provides consistent printing, and typically includes quality control (QC) data.
   An alternative is custom peptide synthesis followed by in-house array printing. In this approach, crude or purified peptides (≥70%–90% purity) can be ordered or synthesized with an N-terminal free amine (for NHS-activated slides) or with C-terminal modifications as required. Including an optional N-terminal spacer (e.g., GGG) can help reduce steric hindrance. This strategy provides complete control over peptide sequences (e.g., enabling alanine scanning) and allows reprints on demand. Peptides are generally spotted directly onto membranes using standard Fmoc chemistry.
   Standard Troubleshooting: When observing faint/variable spots, increase peptide concentration, add 10%–30% glycerol, and adjust humidity (too dry → small/faint spots; too humid → spreading).
   In the case of a high background, improve blocking (try casein vs BSA), include detergent (0.05% Tween-20), and extend the post-print quench (slides).
8. For puromycin selection, there is no need for the killing curve and non-transfected control—no resistant colonies were ever observed in the controls.
9. Only a fraction of cells will be successfully edited when performing CRISPR-mediated genome editing. Depending on the experimental setup, several selection methods are commonly used to enrich or isolate those cells. We have reported a process of selection based on the temperature sensitivity of a cell line that we established [25]. However, other methods were reported, including the antibiotic selection. A selectable marker gene (e.g., puromycin, neomycin, hygromycin resistance) is codelivered with CRISPR components. Cells that used CRISPR/Cas9 machinery (and marker) survive antibiotic treatment. However, the resistant cells are not necessarily edited, so further validation is required. Another method is based on fluorescence-based selection (FACS). In this protocol, a fluorescent protein reporter is coexpressed with Cas9/sgRNA, and FACS sorts cells to enrich for high-expression cells. Still, like antibiotics, it enriches for transfected cells rather than directly edited cells. A more direct protocol is single-cell cloning. Edited populations are diluted and expanded into single-cell clones. The site of editing of each clone is sequenced to ensure the identification of the edited clones [24].

References

1. Collins GA, Goldberg AL (2017) The logic of the 26S proteasome. Cell 169:792–806. https://doi.org/10.1016/j.cell.2017.04.023
2. Zheng N, Shabek N (2017) Ubiquitin ligases: structure, function, and regulation. Annu Rev Biochem 86:129–157. https://doi.org/10.1146/annurev-biochem-060815-014922
3. Asher G, Reuven N, Shaul Y (2006) 20S proteasomes and protein degradation "by default". BioEssays 28:844–849. https://doi.org/10.1002/bies.20447
4. Erales J, Coffino P (2014) Ubiquitin-independent proteasomal degradation. Biochim Biophys Acta 1843:216–221. https://doi.org/10.1016/j.bbamcr.2013.05.008
5. Makaros Y, Raiff A, Timms RT, Wagh AR, Gueta MI, Bekturova A, Guez-Haddad J, Brodsky S, Opatowsky Y, Glickman MH, Elledge SJ, Koren I (2023) Ubiquitin-independent proteasomal degradation driven by C-degron pathways. Mol Cell 83:1921–1935. https://doi.org/10.1016/j.molcel.2023.04.023
6. Zhang Z, Mena EL, Timms RT, Koren I, Elledge SJ (2025) Degrons: defining the rules of protein degradation. Nat Rev Mol Cell Biol 26:868–883. https://doi.org/10.1038/s41580-025-00870-z
7. Asher G, Tsvetkov P, Kahana C, Shaul Y (2005) A mechanism of ubiquitin-independent proteasomal degradation of the tumor suppressors p53 and p73. Genes Dev 19:316–321. https://doi.org/10.1101/gad.319905
8. Adamovich Y, Shlomai A, Tsvetkov P, Umansky KB, Reuven N, Estall JL, Spiegelman BM, Shaul Y (2013) The protein level of PGC-1α, a key metabolic regulator, is controlled by NADH-NQO1. Mol Cell Biol 33:2603–2613. https://doi.org/10.1128/MCB.01672-12
9. Varshavsky A (2024) N-degron pathways. Proc Natl Acad Sci U S A 121:e2408697121. https://doi.org/10.1073/pnas.2408697121
10. Harris TJ, Trader DJ (2025) Exploration of degrons and their ability to mediate targeted protein degradation. RSC Med Chem 16:1067–1082. https://doi.org/10.1039/d4md00787e
11. Asher G, Lotem J, Sachs L, Kahana C, Shaul Y (2002) Mdm-2 and ubiquitin-independent p53 proteasomal degradation regulated by NQO1. Proc Natl Acad Sci U S A 99:13125–13130. https://doi.org/10.1073/pnas.202480499
12. Kalejta RF, Shenk T (2003) Proteasome-dependent, ubiquitin-independent degradation of the Rb family of tumor suppressors by the human cytomegalovirus pp71 protein. Proc Natl Acad Sci U S A 100:3263–3268. https://doi.org/10.1073/pnas.0538058100
13. Tsvetkov P, Asher G, Paz A, Reuven N, Sussman JL, Silman I, Shaul Y (2008) Operational definition of intrinsically unstructured protein sequences based on susceptibility to the 20S proteasome. Proteins 70:1357–1366. https://doi.org/10.1002/prot.21614
14. Tsvetkov P, Shaul Y (2012) Determination of IUP based on susceptibility for degradation by default. Methods Mol Biol 895:3–18. https://doi.org/10.1007/978-1-61779-927-3_1
15. Sahu I, Mali SM, Sulkshane P, Xu C, Rozenberg A, Morag R, Sahoo MP, Singh SK, Ding Z, Wang Y, Day S, Cong Y, Kleifeld O, Brik A, Glickman MH (2021) The 20S as a stand-alone proteasome in cells can degrade the ubiquitin tag. Nat Commun 12:6173. https://doi.org/10.1038/s41467-021-26427-0
16. Myers N, Olender T, Savidor A, Levin Y, Reuven N, Shaul Y (2018) The Disordered landscape of the 20S proteasome substrates reveals tight association with phase separated granules. Proteomics 18:e1800076. https://doi.org/10.1002/pmic.201800076
17. Biran A, Myers N, Steinberger S, Adler J, Riutin M, Broennimann K, Reuven N, Shaul Y (2022) The C-terminus of the PSMA3 proteasome subunit preferentially traps intrinsically disordered proteins for degradation. Cells 11:3231. https://doi.org/10.3390/cells11203231
18. Sahu I, Glickman MH (2021) Structural insights into substrate recognition and processing by the 20S proteasome. Biomolecules 11:148. https://doi.org/10.3390/biom11020148
19. Hu C-D, Grinberg AV, Kerppola TK (2005) Visualization of protein interactions in living cells using bimolecular fluorescence complementation (BiFC) analysis. Curr Protoc Protein Sci Chapter 19:Unit 19.10. https://doi.org/10.1002/0471140864.ps1910s41
20. Touitou R, Richardson J, Bose S, Nakanishi M, Rivett J, Allday MJ (2001) A degradation signal located in the C-terminus of p21WAF1/CIP1 is a binding site for the C8 alpha-subunit of the 20S proteasome. EMBO J 20:2367–2375. https://doi.org/10.1093/emboj/20.10.2367
21. Sdek P, Ying H, Chang DLF, Qiu W, Zheng H, Touitou R, Allday MJ, Xiao Z-XJ (2005) MDM2 promotes proteasome-dependent ubiquitin-independent degradation of retinoblastoma protein. Mol Cell 20:699–708. https://doi.org/10.1016/j.molcel.2005.10.017
22. Riutin M, Erez P, Adler J, Biran A, Myers N, Shaul Y (2024) Investigating the p21 ubiquitin-independent degron reveals a dual degron module regulating p21 degradation and function. Cells 13:1670. https://doi.org/10.3390/cells13191670

23. Reuven N, Shanzer M, Shaul Y (2019) Hippo pathway regulation by tyrosine kinases. Methods Mol Biol 1893:215–236. https://doi.org/10.1007/978-1-4939-8910-2_17
24. Reuven N, Shaul Y (2022) Selecting for CRISPR-edited knock-in cells. Int J Mol Sci 23:11919. https://doi.org/10.3390/ijms231911919
25. Reuven N, Adler J, Broennimann K, Myers N, Shaul Y (2019) Recruitment of DNA repair MRN complex by intrinsically disordered protein domain fused to Cas9 improves efficiency of CRISPR-mediated genome editing. Biomolecules 9:584. https://doi.org/10.3390/biom9100584
26. Reuven N, Adler J, Myers N, Shaul Y (2021) CRISPR co-editing strategy for scarless homology-directed genome editing. Int J Mol Sci 22:3741. https://doi.org/10.3390/ijms22073741
27. Bond SR, Naus CC (2012) RF-cloning.org: an online tool for the design of restriction-free cloning projects. Nucleic Acids Res 40:W209–W213. https://doi.org/10.1093/nar/gks396

Check for updates

Chapter 3

Utilization of Functional Degradation Pathways for the Characterization of Ubiquitin-Proteasome System Tool Compounds

Julia M. Frischkorn, Kristin Riching, Stefan Knapp and Martin P. Schwalm

Abstract

The ubiquitin-proteasome system (UPS) is a highly regulated enzymatic machinery responsible for selective protein degradation, which is an essential process for maintaining cellular proteostasis, signaling, and homeostasis. Dysregulation of the UPS has been implicated in a variety of human diseases, including cancer, neurodegeneration, and inflammatory disorders. Recently, small molecule degraders such as PROTACs and molecular glues emerged that exploit the UPS to selectively degrade disease-associated proteins, and these new modalities are increasingly interesting for drug development and target validation. However, due to the complex multistep pathway, validation of the molecular mechanism of action remains challenging. A number of tool compounds have been developed that can be used to demonstrate that observed PROTAC-induced target degradation relies on the UPS. Despite the availability of these tool compounds, there is no consensus in the community regarding the most appropriate use for these compounds, in particular their dosing and best timepoints used for such functional studies. In this chapter, we present a protocol for the best practice use of UPS-targeting tool compounds to further understand their molecular mechanism of action within the CRBN- and VHL-mediated degradation pathways in human cells. We make use of the HiBiT system supported by a cell viability assay to enable identification of effective, nontoxic concentrations for target degradation rescue experiments. This protocol establishes a framework for performing functional studies and guides the informed, effective use of UPS tool compounds in both research and therapeutic discovery.

Keywords PROTAC, Ubiquitin-proteasome system, HiBiT, CRBN, VHL

1 Introduction

Ubiquitination was discovered approximately 40 years ago and has since then manifested itself as a key post-translational control system that is present in eukaryotic cells [1]. Modern mass spectrometry-based proteomics have already cataloged tens of thousands of ubiquitination sites across numerous proteins, underscoring that most proteins in a cell will carry a ubiquitin mark at some point during their lifetime [1].

Maria W. Górna et al. (eds.), *Degrons*, Methods in Molecular Biology, vol. 3069,
https://doi.org/10.1007/978-1-0716-5508-5_3,

Ubiquitin itself, a protein composed of 76 amino acids, is structurally conserved across eukaryotes, from humans to yeast [2]. Ubiquitin is expressed as polyubiquitin and cleaved to circulate freely in its monomeric form in the cytosol ready to be covalently attached to other proteins, a process called ubiquitination. Ubiquitination follows an ATP-dependent, three-step cascade starting with a ubiquitin-activating enzyme (E1), which hydrolyses ATP to form a high-energy thioester linkage between free ubiquitin (C-terminal carboxyl) and the E1 (cysteine group). The modifier is then transferred to the ubiquitin-conjugating enzyme (E2) through transesterification. The final handoff requires a ubiquitin ligase (E3), which brings the ubiquitin-charged E2 into proximity with a substrate and catalyzes formation of an isopeptide bond between the ubiquitin's C-terminal glycine and the ε-amino group of a lysine from the target protein [1, 2].

While the numbers of E1 and E2 enzymes are small, the E3 ubiquitin ligases form a large and diverse family with over 600 different members in the human proteome [3]. Due to the limited number of E1s and E2s, E3s mark the selectivity determining step in the ubiquitin-proteasome system. They fall broadly into two mechanistic classes: (I) homology to the E6-associated protein carboxyl terminus (HECT)-domain ligases, which form intermediates with ubiquitin before transfer, and (II) Really Interesting New Gene (RING)-domain ligases, which act as scaffolds to position E2 and a substrate. The RING group comprises both single-subunit enzymes, such as Cbl family E3 ligases, and multiprotein complexes, such as Cullin–RING ligases (CRL) [1, 2]. E3 ubiquitin ligases facilitate the last step in the ubiquitin transfer cascade by recruiting a ubiquitin-loaded E2 and a substrate for transfer of the ubiquitin onto an exposed lysine on the target protein's surface. Multiple ubiquitin proteins can be linked via different lysine residues forming a polyubiquitin chain. In case of linkage at lysine 48 (K48), this form of ubiquitination is recognized as a mark for degradation by the cellular proteasome machinery initiating target degradation through the 26S proteasome [2].

The first FDA approval of the proteasome inhibitor bortezomib in 2003 [4] demonstrated that therapeutic intervention within the UPS is feasible, sparking diverse drug discovery campaigns yielding numerous compounds interfering with nearly all steps within the UPS cascade [5]. These steps range from the activation of ubiquitination to the final step of proteasomal degradation, as well as the targeting of regulatory pathways such as neddylation and deubiquitination. Since 2003, several additional drugs targeting the UPS have already been approved by the FDA, which are also useful as chemical tools to study the UPS [6].

Even though not clinically relevant, the proteasome inhibitor MG132 is considered to be of great value for mechanistic studies of proteasome function by inhibiting all proteasome activity and has recently emerged as a standard control compound to show proteasome-dependent degradation in the field of targeted protein degradation (TPD) [7, 8].

Since inhibition of the proteasome targets the last step of the UPS cascade, additional small molecules have been developed to enable specific interrogation of upstream steps [9]. The first step, ubiquitin activation through E1 enzymes such as UAE and UBA6, constitutes an attractive point at which to inhibit ubiquitin activation, and therefore, block the pathway at the initiation step. The UAE is responsible for more than 99% of ubiquitin conjugation, therefore, inhibition causes decrease of a broad spectrum of UPS activity. Mechanistically, the UAE inhibitor forms a covalent adduct with ubiquitin (Ub) displaying high affinity for the UAE active site [10]. As a broad-spectrum inhibitor for ubiquitin activation, TAK-243 has been developed [10].

Sharing structural similarity with ubiquitin, Neural precursor cell Expressed Developmentally Down-regulated protein 8 (NEDD8) marks another ubiquitin-like modifier involved in the human UPS. Unlike ubiquitin, transfer of NEDD8 to the Cullin–RING ligase (CRL) subset of E3 ligases is required for their activity, which requires an E1, E2, E3 cascade as well. Therefore, the NEDD8-activating enzyme (NAE) has emerged as a NEDD8 selective E1 equivalent, regulating CRL activity within the UPS, sparing all residual UPS activities not involving this subset of E3 ligases [11]. The most advanced so called neddylation inhibitor, MLN-4924 (Pevonedistat), targets NAE through a similar pathway as ubiquitin E1 inhibitors by forming a drug-NEDD8 adduct, blocking NAE function. Blocking neddylation and inhibiting CRL activation was found to induce the accumulation of CRL substrates, including p21, p27, and Wee1 [12].

As a counter regulation, deneddylation marks the cellular process of removing NEDD8 from target proteins, primarily regulated by the COP9 signalosome [13]. The proteolytic activity resides in the CSN5 subunit, which contains the catalytic metalloprotease domain [14]. This process is essential for cullin-RING ligase regulation, as cycles of neddylation and deneddylation control ligase activation, substrate receptor exchange, and protein degradation [15]. The small molecule CSN5i-3 is an inhibitor of CSN5 and was developed to prevent deneddylation [14, 16].

The process of ubiquitination is reversible through deubiquitinases (DUBs), which can remove ubiquitin from substrates or edit chains, returning proteins to their unmodified state and recycling ubiquitin for the introduction of new Ub marks [2]. Inhibition of DUBs was found to destabilize proteins by interfering with the cellular process of removal of the polyubiquitin chain. In contrast, protein stabilization of target proteins marked for degradation can be achieved by inhibition of the UPS cascade [17].

Due to the importance of E3 ligases in recruiting substrates to the UPS, this large and diverse protein family also became an attractive target for small molecule development apart from modulation of their activity state using, e.g., neddylation inhibitors [5]. Nevertheless, rational design of E3 ligase binders turned out to be challenging. A recent success story originating from the Ciulli lab describes the development of VH-298, a chemical probe for the VHL (von Hippel-Lindau) E3 ligase playing a crucial role in hypoxia signaling through HIF-1α (hypoxia-inducible factor-1α) regulation [18]. Other molecules such as lenalidomide and thalidomide were not discovered through rational design and only later found to be ligands for an E3 ligase. Lenalidomide binds to cereblon (CRBN) and acts as a molecular glue that reprograms the ligase's substrate specificity to recruit the B-cell transcription factors IKZF1 (Ikaros) and IKZF3 (Aiolos) for ubiquitination and proteasomal degradation [19]. These examples highlight the importance of small molecules interfering with the UPS ranging from tool compounds to clinical applications [7].

Since disruption of this tight regulatory network can have far-reaching consequences, the UPS has become an attractive target for therapeutic intervention. Moreover, recent advances in TPD are actively exploiting the UPS, and clinical degrader molecules have emerged as a new and transformative approach in drug discovery and chemical biology [7]. For developing TPD systems, small molecules are used to induce proximity between an E3 ligase and a non-native target protein substrate, initiating polyubiquitination and subsequent proteasomal degradation of the target. Mostly, these small molecules are part of either the group of bispecific PROteolysis TArgeting Chimeras (PROTACs) or molecular glues. While PROTACs are still currently undergoing clinical trials, molecular glues such as thalidomide have already been approved by the FDA decades even before their mechanism of action was deciphered [19]. Moreover, having E3 ligase ligands in hand, they can be utilized for the development of PROTACs to induce protein degradation of non-native substrates [20]. This is possible by chemically connecting two ligands, each specific for a target protein of interest and an E3 ligase component, with a linker moiety. Upon binding of the PROTAC to both target and E3 ligase, a ternary complex is formed, leading to the ubiquitination of the protein of interest (POI) and subsequent degradation through the proteasome system (Fig. 1) [7].

The high complexity of the multistep pathway complicates the generalized UPS inhibitor measurements due to the different steps at which the inhibition can occur. Therefore, one of the few possibilities is the measurement of the UPS inhibitor's influence on target degradation. In contrast to PROTAC screening where the individual steps can be monitored, this approach generalizes the UPS as one

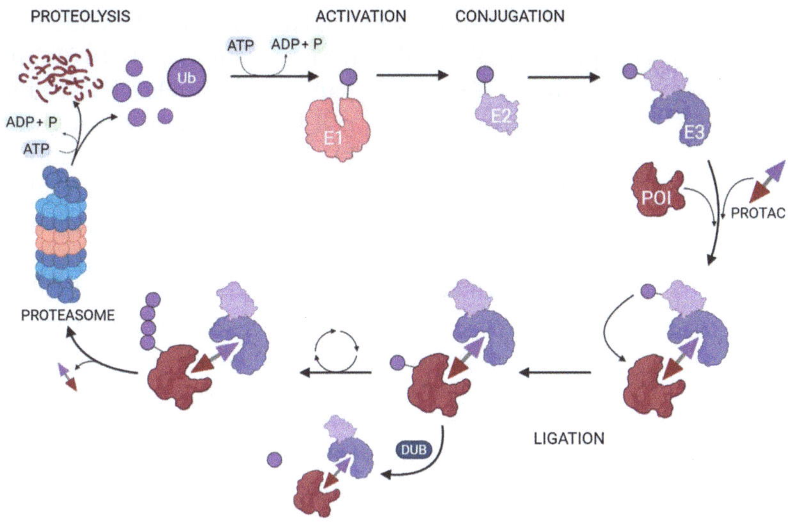

Fig. 1 Schematic representation of small molecule-induced targeted protein degradation. Ubiquitin is activated through an E1 enzyme followed by transfer onto an E2. In this example, the Ub-charged E2 is loaded onto a RING E3 ligase. The protein of interest is recruited by a heterobifunctional small molecule to the E3, bringing it into close proximity, now called neosubstrate (or non-natural substrate). This close proximity and optimal presentation of a surface-exposed lysine results in Ub transfer onto the protein of interest (POI). After multiple cycles, the polyubiquitinated POI is recognized by the 26S proteasome causing POI degradation and release of Ub for another cycle. The small molecule typically dissociates and is reused

unit. Exemplary individual steps in PROTAC characterization that can be specifically studied are ternary complex formation, ubiquitination efficiency and degradation [21]. Western blot experiments are often used to evaluate target degradation as the final step of the cascade through abundance determination. However, confirmation by further experiments is often required to extract reliable potency data due to limited dynamic range of the method [22]. To assess broader mechanistic insights into the mode-of-action of the compound, approaches such as quantitative live cell assays can be used [21]. To monitor protein abundance and its changes in living cells, proteins can be endogenously tagged with an 11 amino acid peptide tag called HiBiT [23]. HiBiT is an engineered peptide derived from a split NanoLuc luciferase (NLuc), in which the larger subunit is called LgBiT. The high affinity of HiBiT and LgBiT results in spontaneous complementation and formation of a bright NanoBiT luciferase with light output proportional to the protein's endogenously regulated abundance. The small tag and brightness of the complemented NanoBiT luciferase allow preservation of the protein's endogenous context, while providing a high dynamic range [23]. Past studies have demonstrated the advantages of using such live cell assays to confirm that ternary complex efficiency correlates with degradation potency in cellular contexts [8]. In addition, proteasome rescue experiments validate whether a degradation is UPS-dependent [7]. Furthermore, pharmacological parameters such as half-maximum degradation concentration (DC_{50}) and maximum level of degradation (D_{max}) can be determined with cell-based systems through monitoring target protein abundance [21, 22]. These aspects combined show that live cell assays show a sufficient dynamic range to reliably determine DC_{50} and D_{max}, track target protein abundance and allow to see effects from the endogenous UPS machinery [8, 21].

The development and optimization of PROTAC molecules heavily rely on chemical tool compounds that selectively inhibit distinct steps of the UPS pathway [7, 8]. These control compounds serve as validation tools to demonstrate that protein degradation is indeed proteasome-dependent and occurs through the expected mechanistic pathway. Despite their widespread use in the literature, comprehensive characterization of these pathway inhibitors has been notably lacking. This especially includes their optimal working concentrations, cellular toxicity profiles, and potential off-target effects [7]. Here, we describe a protocol that addresses this gap in the drug development workflow by providing a method for characterizing UPS pathway inhibitors (Fig. 2). To address this, well studied PROTACs can be used as a chemical tool compound to also study interference with the UPS, which is reflected in changes in target protein abundance [24, 25]. The methodology focuses on two well-established E3 ligase systems: (I) cereblon (CRBN) and (II) von Hippel-Lindau (VHL). However, the degradation workflow can be applied to any degrader including those that recruit novel E3 ligases [7]. Therefore, in the first step, the PROTACs must be studied to obtain the concentration at which the degradation effect is maximum (DC_{100}), which later reflects the assay window (schematic workflow depicted in Fig. 3 with a kinetic workflow shown in Fig. 4). In the next step, this concentration is used to chemically knock-down a target protein and study dose-dependent effects of compounds, which interfere with the UPS machinery (workflow depicted in Fig. 3). A last step marks the study of compound effects on cell viability, which

Fig. 2 Assay cascade for characterizing UPS pathway inhibitors. Degradation (left): PROTAC-driven loss of protein of interest (POI) is quantified across dose and time. Degradation rescue (middle): At the established DC_{100} (induced POI knockdown), live cell rescue experiments assess UPS dependence and rank inhibitor potency by titrating pathway inhibitors. As an optional step, orthogonal confirmation by western blot enables cross-cell line validation. Viability (right): Cytotoxicity readout is run in parallel to separate genuine pathway-specific rescue from nonspecific changes in signal due to cell health effects and to define non-toxic concentration ranges

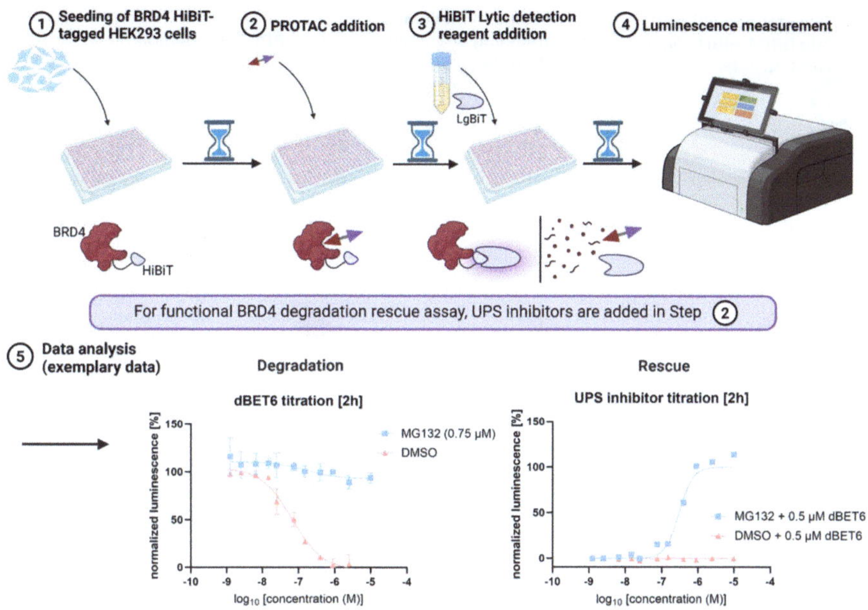

Fig. 3 Workflow for HiBiT end point degradation and rescue assays. (1) $BRD4^{HiBiT}$ cells are plated into small-volume 384-well plates. (2) For degradation, an 11-point PROTAC dose-response is added. For the degradation rescue format, UPS inhibitors are titrated while the degrader is held constant at the DC_{100}. (3) Nano-Glo HiBiT Lytic detection reagent is added to lyse cells and reconstitute luciferase. (4) Luminescence is measured on a plate reader. (5) The data are fitted using a normalized three-parameter curve as shown in the exemplary data generated with dBET6 and MG132. Degradation is shown by a decreasing dose–response curve (left), whereas rescue causes recovery of signal with increasing inhibitor concentration (right). As seen on the left, the degradation effect can be suppressed by addition of UPS inhibitors

was found to often contribute to misleading data while measuring protein abundance as both degrader and UPS pathway modulators can exhibit time- and dose-dependent cytotoxic effects (workflow depicted in Fig. 5) [7]. Optionally, the degradation and its rescue can be confirmed by western blot. In brief, this protocol will cover the steps outlined in Fig. 2, which can also be executed individually. In case of novel PROTACs or molecular glues, it is recommended to include CellTiter-Glo measurements to ensure the absence of toxic effects when determining degradation and rescue. If toxic effects are observed, these off-target effects have to be taken into account when analyzing the data. For rescue, we recommend using the lowest concentration and shortest timepoint that yields the greatest degradation, unless cytotoxic effects are observed. In this case, a shorter timepoint or even lower concentration should be considered. This is especially important when cytotoxic effects may arise from both the PROTAC and the UPS inhibitor.

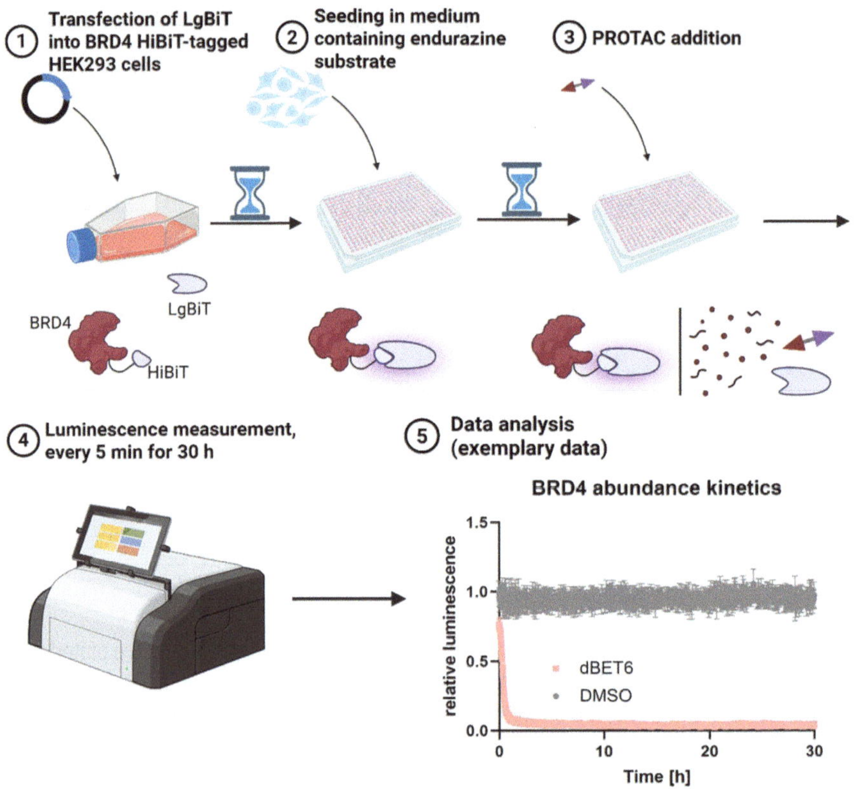

Fig. 4 Workflow for kinetic HiBiT degradation assays. (1) Transfection of BRD4^{HiBiT} cells with a LgBiT expression vector to enable intracellular HiBiT–LgBiT complementation (≥16 h expression). (2) Transfected BRD4^{HiBiT} cells are plated into small-volume 384-well plates using CO_2-independent medium supplemented with Nano-Glo endurazine and allowed to equilibrate for 2.5 h. (3) Immediately before measurement, the degrader (or control) is added at the desired concentrations, and plates are sealed. (4) Plates are transferred to a temperature-controlled plate reader (30°C or 37°C), and luminescence is measured at regular intervals (e.g., every 5 min for up to 30 h). The resulting time course reports real-time detection of POI levels. (5) The measured luminescence units are plotted against time as shown in the exemplary data generated with dBET6 and a DMSO control. Degradation is expressed by a decreasing curve

This investigation is particularly valuable given that many inhibitor concentrations commonly reported in the literature fall within toxic ranges, potentially leading to artifact-biased experimental outcomes. It ensures that researchers can distinguish genuine pathway-specific effects from non-specific cellular disruption, ultimately improving the quality and reproducibility of targeted protein degradation research [7]. A workflow validating new E3 ligands for PROTAC development that also considers cell toxicity has recently been established [26].

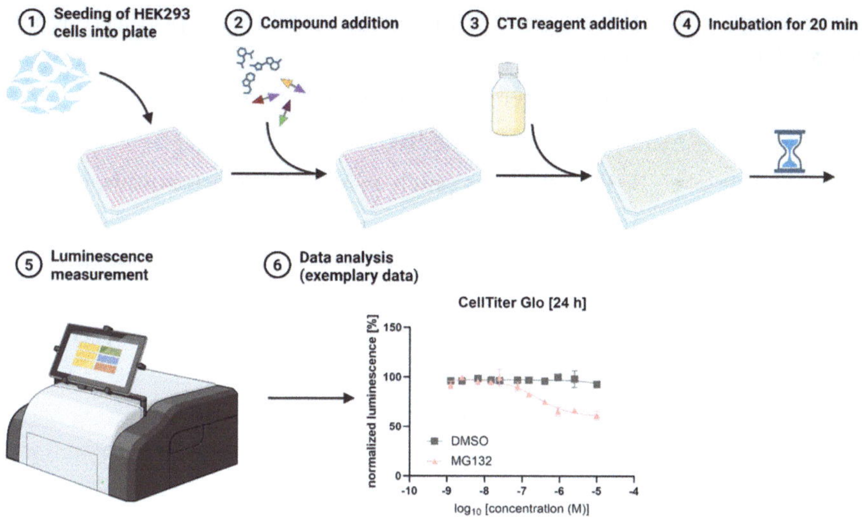

Fig. 5 Viability determination using the CellTiter-Glo assay. (1) HEK293T cells are seeded in white small-volume 384-well plates and allowed to attach (approx. 24 h). (2) PROTACs and UPS inhibitors (11-point dose–response; vehicle controls included) are added and plates are briefly centrifuged; cells are incubated for the desired time. (3) An equal volume of CellTiter-Glo 2.0 (CTG) reagent is dispensed to each well. (4) The plate is incubated for 20 min in the dark. (5) Luminescence is measured on a plate reader. (6) The normalized luminescence units are plotted against the logarithmized concentration as shown in the exemplary data generated with MG132 and a DMSO control. A decrease in cell viability is indicated by a decrease in the luminescence signal

2 Materials

2.1 Cell Maintenance

- Endogenously BRD4 HiBiT-tagged HEK293T cells ($BRD4^{HiBiT}$) (Promega, CS3023270).
- HEK293T cells.
- Dulbecco's Modified Eagle Medium (DMEM).
- Fetal bovine serum (FBS).
- T75 flask.
- Dulbecco's phosphate-buffered saline (DPBS).
- Trypsin-EDTA (1×), 0.25%.

2.2 HiBiT End Point Detection for BRD4 Degradation and DC_{100} Determination

- White small volume 384-well plate, polystyrene surface, flat bottom.
- Plate reader for luminescence detection.
- BRD4 degrader (dBET6 or MZ1).
- DMSO.
- Nano-Glo HiBiT Lytic detection reagent (Promega, N3040).

2.3 Functional Assay for BRD4 Degradation Rescue

- White small volume 384- well plate, polystyrene surface, flat bottom.
- Plate reader for luminescence detection.
- BRD4 degrader (dBET6 or MZ1).
- UPS inhibitors (such as MG132, bortezomib, or MLN4924).
- DMSO.
- Nano-Glo HiBiT Lytic detection reagent (Promega, N3040).

2.4 BRD4 Degradation Kinetics

- White small volume 384-well plate, polystyrene surface, flat bottom.
- Plate reader for luminescence detection.
- BRD4 degrader (dBET6 or MZ1).
- DMSO.
- OptiMEM phenol red- and serum-free medium.
- FuGENE 4K.
- LgBiT expression vector.
- Transfection Carrier DNA.
- T225 flask.
- CO_2-independent medium.
- Nano-Glo Endurazine substrate (Promega, N2570).
- Breathable plate seal.

2.5 CellTiter-Glo Viability Determination

- CellTiter-Glo 2.0 Cell Viability Assay (Promega, G9241).
- White small volume 384 well plate, polystyrene surface, flat bottom.
- DMSO.
- Small molecule inhibitors of interest.

2.6 Western Blot

- 6-well tissue culture plate.
- BRD4 degrader (dBET6 or MZ1).
- RIPA buffer: 150 mM NaCl, 1% Triton X-100, 0.1% Sodium Dodecyl Sulfate (SDS), 50 mM Tris Hydrochloride (pH = 8.0).
- 6× Laemmli buffer: 15 mL 1 M Tris Hydrochloride (pH = 6.8), 6 g Sodium Dodecyl Sulfate (SDS), 0.03% Bromophenol blue, 30 mL Glycerol, 2 mL 14.3 M β-mercaptoethanol.
- Bicinchoninic acid (BCA) protein assay.
- Protein Ladder.
- PVDF membrane.
- SDS-PAGE gel.

- TBST: 100 mL 10× TBS (pH 7.6), 900 mL ddH_2O, 1 mL Tween-20.
- Transfer buffer: 25 mM Tris Base (pH = 8.3), 190 mM glycine, 20% methanol.
- Running buffer: 25 mM Tris Base (pH = 8.3), 190 mM glycine, 0.1% Sodium Dodecyl Sulfate (SDS).
- 5% (wt/vol) bovine serum albumin (BSA).
- Primary antibody BRD4.
- Primary antibody internal standard.
- HRP-conjugated secondary antibodies.
- HRP substrate.
- Nano-Glo HiBiT Blotting System (Promega, N2410).

3 Methods

3.1 HiBiT End Point Detection for BRD4 Degradation and DC_{100} Determination

1. Cultivate endogenously BRD4 HiBiT-tagged HEK293T ($BRD4^{HiBiT}$) cells in DMEM cell culture medium (supplemented with 10% FBS) in T75 flasks (10–15 mL) to a confluency of 70%–90% (*see* **Note 1**).
2. Aspirate the culture medium carefully without aspirating the cells.
3. Briefly wash the cells by adding sterile, prewarmed 5 mL DPBS to the cells (*see* **Note 2**).
4. Carefully aspirate the DPBS.
5. Add 2.5 mL Trypsin onto the cells and incubate at 37°C and 5% CO_2 until detachment of the cells (approx. 5 min).
6. After detachment, stop trypsinization by addition of 7.5 mL prewarmed DMEM cell culture medium supplemented with 10% FBS.
7. To change the medium, pellet the cells by centrifugation at 200 ×*g* for 5 min.
8. Carefully aspirate the medium and add 10 mL fresh prewarmed DMEM cell culture medium (supplemented with 10% FBS).
9. Count the cells, e.g., by using an automated cell counter (*see* **Note 3**).
10. Prepare a cell suspension in prewarmed DMEM medium (supplemented with 10% FBS) containing 2.5×10^5 cells/mL calculating 4 mL cells per plate (with the addition of an appropriate dead volume).

11. Seed 10 μL of this suspension into each well of a small volume 384-well plate except for column 24, which is filled with DMEM medium (no cell control).
12. Add PROTACs in dose-response using 11 points for high resolution. An initial titration might be from 10 to 0.001 μM. Using 11 points allows for technical replicates in a single row (including two negative controls in column 12 and 24).
13. Centrifuge plate at 200 ×*g* for 1 min.
14. Incubate the plate of treated cells at 37°C and 5% CO_2 for the desired time (*see* **Note 4**).
15. After incubation, prepare the Nano-Glo HiBiT Lytic detection reagent according to the manufacturer's protocol by diluting the LgBiT protein 1:50 and the Lytic substrate 1:100 in the Lytic buffer (equilibrated to room temperature). Prepare 4 mL per plate (*see* **Note 5**).
16. Add 10 μL of the prepared Lytic detection solution to each well and incubate the plate in the dark for 10 min at room temperature.
17. Centrifuge plate at 200 ×*g* for 1 min.
18. Measure luminescence in a compatible plate reader such as a BMG PHERAstar FSX or GloMax Discover.
19. Plot the obtained luminescent units against the degrader concentration and normalize against the negative control (column 12 = 100%) and the no cell control (column 24 = 0%).
20. The data can be fitted using a normalized three-parameter curve fit with the following equation:

$$Y = \frac{100}{\left(1 + 10^{(X - LogIC_{50})}\right)}$$

21. Determine the lowest degrader concentration necessary for complete degradation (DC_{100}).

3.2 Functional Assay for BRD4 Degradation Rescue

1. Cultivate BRD4^{HiBiT} cells in DMEM cell culture medium (supplemented with 10% FBS) (*see* **Note 1**) and prepare an assay-ready 384-well plate by following Steps 1–11 of the DC_{100} determination protocol. In this protocol, there is no requirement for a no cell control.
2. In this experiment, the UPS inhibitor of interest is titrated while keeping the degrader concentration constant at the DC_{100} determined in Step 21. Keep the concentration range identical but pay attention to not exceeding a total of 1% DMSO when adding degrader and UPS inhibitor. In case of dBET6 500 nM is required for full degradation, in the case

of MZ1 1 μM is required. Add the degrader in column 1–23 and the titration of the UPS inhibitors in column 1–11 and 13–23.

3. Centrifuge plate at 200 × *g* for 1 min.
4. Incubate the plate for the time of interest at 37°C and 5% CO_2 (*see* **Note 4**).
5. After incubation, prepare the Nano-Glo HiBiT Lytic detection reagent according to the manufacturer's protocol by diluting the LgBiT protein 1:50 and the Lytic substrate 1:100 in the Lytic buffer (equilibrated to room temperature). Prepare 4 mL per plate with additional dead volume.
6. Add 10 μL of the prepared Lytic detection solution to each well and centrifuge plate at 200 × *g* for 1 min.
7. Incubate the plate in the dark for 10 min at room temperature.
8. Measure luminescence in a compatible plate reader such as a BMG PHERAstar FSX.
9. Plot the obtained luminescent units against the inhibitor concentration and normalize against the full degradation control (column 12 = 0%) and the no degrader control (column 24 = 100%).
10. The data can be fitted using a normalized three-parameter curve to fit with the following equation for IC_{50} determination of the UPS inhibitors.

$$Y = \frac{100}{\left(1 + 10^{(X - LogIC_{50})}\right)}$$

3.3 BRD4 Degradation Kinetics

1. Cultivate BRD4^{HiBiT} cells in DMEM cell culture medium (supplemented with 10% FBS) and prepare a counted cell suspension by following Steps 1–7 of the DC_{100} determination protocol (*see* **Note 1**).
2. Prepare LgBiT transfection complexes by mixing 1 μg LgBiT expression vector DNA with 9 μg transfection carrier DNA in 1 mL OptiMEM phenol red- and serum-free medium in a 1.5 mL reaction tube.
3. Add 30 μL/mL FuGENE 4 K and mix by inverting the tube five times (*see* **Note 6**).
4. Incubate the transfection mix for 30 min at room temperature.
5. Prepare 20 mL of the counted cell solution (*see* **Note 3**) in a 50 mL tube and slowly add the formed transfection complexes to the solution.
6. Invert the tube five times.

7. Transfer the cell suspension to a T225 flask and incubate for a minimum of 16 h at 37°C and 5% CO_2 to express the LgBiT protein (*see* **Note 7**).
8. Harvest the cells by following Steps 2–6 from the DC_{100} determination protocol.
9. Resuspend the cells in 5 mL prewarmed CO_2-independent medium containing Nano-Glo endurazine substrate (*see* **Note 8**).
10. Seed 10 μL of the solution into each well of white small volume 384 well plates.
11. Incubate the plate for 2.5 h at 37°C and 5% CO_2.
12. For readout, preheat the plate reader to 37°C or 30°C if possible (*see* **Note 9**).
13. Immediately before readout, add the compound of interest in the concentration of interest to the wells and seal the plate using a breathable seal.
14. Centrifuge plate at 200 × *g* for 1 min.
15. Transfer the plate to the preheated plate reader and continuously measure luminescence every 5 min for 30 h in a plate reader.
16. Data can be graphed using the measured luminescence units plotted against time.

3.4 CellTiter-Glo Viability Determination

1. Cultivate HEK293T cells in DMEM cell culture medium (supplemented with 10% FBS) and prepare a counted cell suspension by following Steps 1–6 of the DC_{100} determination protocol (*see* **Note 1**).
2. Prepare a cell suspension in prewarmed DMEM medium (supplemented with 10% FBS) containing 1×10^6 cells/mL calculating 4 mL cells per plate (*see* **Note 5**).
3. Seed 10 μL of the prepared HEK293T cell suspension into each well of a white small volume 384 plate, leaving column 24 without cells, containing only DMEM cell culture medium.
4. Incubate the plate for 24 h at 37°C and 5% CO_2.
5. The next day, add PROTACs and UPS inhibitors in dose-response using 11 points for high resolution. An initial titration might be from 10 to 0.001 μM. Using 11 points allows for technical replicates in a single row (excluding columns 12 and 24 as controls) (*see* **Note 10**).
6. Centrifuge plate at 200 × *g* for 2 min.
7. Incubate the plate to desired timepoint at 37°C and 5% CO_2 (for e.g., 2, 6, or 24 h).

8. For measurement, add 10 μL of assay reagent and centrifuge plate at 200 × *g* for 2 min.
9. Incubate in the dark for 20 min at room temperature.
10. Measure luminescence in a compatible plate reader such as a BMG PHERAstar FSX or GloMax Discover.
11. The signal reflects the number of viable cells and can be normalized to vehicle (column 12) and no-cell/background control (column 24).

3.5 Western Blot

1. Cultivate HEK293T cells (or cell line of interest) in the respective cell culture medium (supplemented with 10% FBS) and prepare a counted cell suspension by following Steps 1–7 of the DC_{100} determination protocol (*see* **Note 1**).
2. Seed cells with a total concentration of 0.5×10^6 cells/mL in cell culture medium (supplemented with 10% FBS) into a 6-well tissue culture plate and incubate overnight.
3. Treat the cells with the respective rescue compounds at their recommended concentration, both with and without a degrader at its DC_{100} concentration.
4. Incubate the plate of treated cells at 37°C and 5% CO_2 for the desired time (*see* **Note 4**).
5. Aspirate the cell culture medium from each well and wash the cells with 1 mL of ice-cold DPBS by carefully adding and aspirating.
6. Add 100 μL of ice-cold RIPA buffer in each well.
7. After 1 min, scrape off the lysate while keeping the plate on ice and transfer it into individual 1.5 mL reaction tubes.
8. Keep the reaction tubes on ice for 25 min.
9. Centrifuge the tubes afterwards for 30 min at 15,000 × *g* at 4°C.
10. Transfer the supernatant into fresh reaction tubes.
11. Prepare 50 μL of each sample (containing equal concentrations) (*see* **Note 11**).
12. Add 10 μL of 6× Laemmli buffer to each sample.
13. Heat each sample for 10 min at 70°C.
14. Load 20 μL of each sample and 10 μL of a suitable protein ladder (*see* **Note 12**) onto a precasted SDS-PAGE gel.
15. Run the gel 2 h at 100 V in running buffer.
16. Transfer the samples to a PVDF membrane through wet blotting in transfer buffer (4°C for 90 min at 95 V and 200 mA).
17. Block the membrane for 1 h using TBST supplemented with 5% BSA at room temperature.

18. Wash the membrane three times by shaking for 5 min in TBST.
19. Incubate the membrane overnight at 4°C with the primary antibody against the POI (e.g., BRD4) in TBST containing 5% BSA under constant agitation.
20. After incubation, wash the membrane three times by shaking it for 5 min in TBST.
21. Incubate the membrane with the secondary HRP-conjugated antibody (diluted in TBST containing 5% BSA) for 1 h at room temperature under constant agitation (*see* **Note 13**).
22. Detection of the membrane can be performed in a chemiluminescence imager after addition of HRP substrate according to the manufacturer's protocol.
23. After detection, wash the membrane three times by shaking for 5 min in TBST.
24. Add a primary antibody for an internal standard (e.g., Vinculin or GAPDH) diluted in TBST containing 5% BSA.
25. Incubate overnight at 4°C under constant agitation.
26. After incubation, wash the membrane three times by shaking for 5 min in TBST.
27. Add the secondary HRP-conjugated antibody (diluted in TBST containing 5% BSA) and incubate for 1 h at room temperature with constant agitation (*see* **Note 13**).
28. Detection of the membrane can be performed in a chemiluminescence imager after addition of HRP substrate according to the manufacturer's protocol.

3.6 Western Blot for HiBiT Detection

1. Cultivate BRD4^{HiBiT} cells (or other endogenously tagged cells of interest) in the respective cell culture medium (supplemented with 10% FBS) and prepare a counted cell suspension by following Steps 1–7 of the DC_{100} determination protocol (*see* **Note 1**).
2. Prepare the membrane as described in Steps 2–18 of Sect. 3.5 Western Blot protocol.
3. Dilute LgBiT protein (1:200) in 1 mL Nano-Glo blotting buffer and 9 mL water (*see* **Note 14**).
4. Incubate the membrane in the prepared mix overnight at 4°C while shaking.
5. The next day, equilibrate the membrane to room temperature.
6. Add the Nano-Glo Luciferase assay substrate in a 1:500 dilution.
7. Incubate the substrate on the membrane for 5 min while gently rocking at room temperature.

8. Bioluminescence detection of the membrane can be performed in a chemiluminescence imager.
9. For detection of an internal standard (e.g., Vinculin or GAPDH), wash the membrane three times for 5 min in TBST while rocking and follow the Steps 24–28 described in Sect. 3.5 Western Blot protocol.

4 Notes

1. Penicillin (100 U/mL)/Streptomycin (100 μg/mL) can be added to the cell culture medium to reduce bacterial contamination during routine handling.
2. The DPBS should be added to the side of the flask and allowed to flow slowly over the cells without shaking vigorously to avoid unwanted detachment of the cells.
3. Cells can be counted by, e.g., using trypan blue (0.4%). To do this, mix 10 μL of cell suspension with 10 μL of trypan blue. This solution is placed on a counting slide. Count unstained (live) and blue (dead) cells in the prescribed grid or use an automated cell counter.
4. For unknown degraders, it is recommended to test a shorter (2–6 h) and a longer (24 h) timepoint since the degrader kinetics can vary greatly. For rescue, it is recommended to choose a timepoint that allows for detection at the determined maximum level of degradation.
5. For the preparation of 4 mL for one plate, we recommend preparing additional 1 mL to compensate for dead volume.
6. Add FuGENE 4K directly into the liquid, without touching the tube. Dispense the reagent with the pipette tip submerged in the medium so it mixes instantly. If FuGENE 4K is pipetted on dry plastic, it can adsorb/coat the plastic and bead as a thin film, leading to loss of transfection efficiency. Use low-retention tips/tubes and avoid vortexing (gentle flick/invert instead).
7. Do not leave the transfection mix on cells for longer than 24 h.
8. Prepare a 1× working solution by adding 50 μL of 100× stock to 5 mL medium; if drugs/vehicle will be added later, reduce the starting volume so the final mix remains 1×. Add the mix to the cells and incubate 2.5 h at 37°C before reading to allow uptake/activation. Shorter times often give little or no luminescence. Alternatively, vivazine can be used with other cell-lines using the same 1× dilution and 2 h incubation.

9. Use 37°C for the most physiological condition. Use 30°C for long kinetic reads to reduce condensation (on lids/optics) and evaporation, keep the entire experiment at one temperature once started.
10. Prepare a concentrated drug stock and make a stepwise serial dilution (e.g., 1:2 or 1:3) across tubes using fresh tips and thorough mixing. Predispense assay medium and add equal volumes from each dilution so the final DMSO (vehicle) % is constant across all wells. Include vehicle-only (DMSO) negative controls for column 12. These should match the highest DMSO concentration used.
11. To determine the protein concentration a bicinchoninic acid (BCA) protein assay can be performed. Use BSA standards in a range from 0–2 mg/mL to generate a linear standard curve (absorbance vs. BSA concentration). Use the fit equation to calculate each sample's concentration, then dilute higher-concentration samples to the level of the lowest so all samples have an equal concentration.
12. Protein ladder must be chosen based on target molecular weight.
13. Secondary antibodies must be directed against the species of the primary antibodies.
14. As an alternative to LgBiT detection, anti-HiBiT antibodies can be utilized as primary antibodies using detection via conjugated secondary antibodies.

Acknowledgments

S.K. is grateful for support by the Structural Genomics Consortium (SGC), a registered charity (No. 1097737) that receives funds from Bayer AG, Boehringer Ingelheim, Bristol Myers Squibb, Genentech, Genome Canada through Ontario Genomics Institute [OGI-196], EU/EFPIA/OICR/McGill/KTH/Diamond Innovative Medicines Initiative 2 Joint Undertaking [EUbOPEN Grant 875510], Janssen, Merck KGaA (aka EMD in Canada and U.S.), Pfizer, and Takeda. S.K. acknowledges funding from the German Cancer Consortium (DKTK) at the German Cancer Research Center (DKFZ). Figures were created using BioRender.com.

References

1. Damgaard RB (2021) The ubiquitin system: from cell signalling to disease biology and new therapeutic opportunities. Cell Death Differ 28(2):423–426. https://doi.org/10.1038/s41418-020-00703-w
2. Guo HJ, Rahimi N, Tadi P (2025) Biochemistry, ubiquitination. In: StatPearls. Treasure Island (FL)
3. Shah Zaib Saleem R, Schwalm MP, Knapp S (2024) Expanding the ligand spaces for E3

ligases for the design of protein degraders. Bioorg Med Chem 105:117718. https://doi.org/10.1016/j.bmc.2024.117718

4. Teicher BA, Tomaszewski JE (2015) Proteasome inhibitors. Biochem Pharmacol 96(1):1–9. https://doi.org/10.1016/j.bcp.2015.04.008
5. Deng L et al. (2020) The role of ubiquitination in tumorigenesis and targeted drug discovery. Signal Transduct Target Ther 5(1):11. https://doi.org/10.1038/s41392-020-0107-0
6. Spano D, Catara G (2023) Targeting the ubiquitin-proteasome system and recent advances in cancer therapy. Cells 13(1):29. https://doi.org/10.3390/cells13010029
7. Schwalm MP et al. (2025) Functional characterization of pathway inhibitors for the ubiquitin-proteasome system (UPS) as tool compounds for CRBN and VHL-mediated targeted protein degradation. ACS Chem Biol 20(1):94–104. https://doi.org/10.1021/acschembio.4c00450
8. Schwalm MP et al. (2023) Tracking the PROTAC degradation pathway in living cells highlights the importance of ternary complex measurement for PROTAC optimization. Cell Chem Biol 30(7):753–765. https://doi.org/10.1016/j.chembiol.2023.06.002
9. Nalepa G, Rolfe M, Harper JW (2006) Drug discovery in the ubiquitin-proteasome system. Nat Rev Drug Discov 5(7):596–613. https://doi.org/10.1038/nrd2056
10. Hyer ML et al. (2018) A small-molecule inhibitor of the ubiquitin activating enzyme for cancer treatment. Nat Med 24(2):186–193. https://doi.org/10.1038/nm.4474
11. Lydeard JR, Schulman BA, Harper JW (2013) Building and remodelling Cullin-RING E3 ubiquitin ligases. EMBO Rep 14(12):1050–1061. https://doi.org/10.1038/embor.2013.173
12. Tong S et al. (2017) MLN4924 (Pevonedistat), a protein neddylation inhibitor, suppresses proliferation and migration of human clear cell renal cell carcinoma. Sci Rep 7(1):5599. https://doi.org/10.1038/s41598-017-06098-y
13. Kandala S, Kim IM, Su H (2014) Neddylation and deneddylation in cardiac biology. Am J Cardiovasc Dis 4(4):140–158
14. Schlierf A et al. (2016) Targeted inhibition of the COP9 signalosome for treatment of cancer. Nat Commun 7: 13166. https://doi.org/10.1038/ncomms13166
15. Lin H et al. (2020) Basis for metabolite-dependent Cullin-RING ligase deneddylation by the COP9 signalosome. Proc Natl Acad Sci U S A 117(8):4117–4124. https://doi.org/10.1073/pnas.1911998117
16. Hanzl A et al. (2023) E3-specific degrader discovery by dynamic tracing of substrate receptor abundance. J Am Chem Soc 145(2):1176–1184. https://doi.org/10.1021/jacs.2c10784
17. Lamberto I et al. (2017) Structure-guided development of a potent and selective non-covalent active-site inhibitor of USP7. Cell Chem Biol 24(12):1490–1500. https://doi.org/10.1016/j.chembiol.2017.09.003
18. Frost J et al. (2016) Potent and selective chemical probe of hypoxic signalling downstream of HIF-alpha hydroxylation via VHL inhibition. Nat Commun 7: 13312. https://doi.org/10.1038/ncomms13312
19. Yamamoto J, Ito T, Yamaguchi Y, Handa H (2022) Discovery of CRBN as a target of thalidomide: a breakthrough for progress in the development of protein degraders. Chem Soc Rev 51(15):6234–6250. https://doi.org/10.1039/d2cs00116k
20. Ishida T, Ciulli A (2021) E3 ligase ligands for PROTACs: how they were found and how to discover new ones. SLAS Discov 26(4):484–502. https://doi.org/10.1177/2472555220965528
21. Riching KM et al. (2018) Quantitative live-cell kinetic degradation and mechanistic profiling of PROTAC mode of action. ACS Chem Biol 13(9):2758–2770. https://doi.org/10.1021/acschembio.8b00692
22. Hartung IV et al. (2023) Expanding chemical probe space: quality criteria for covalent and degrader probes. J Med Chem 66(14):9297–9312. https://doi.org/10.1021/acs.jmedchem.3c00550
23. Schwinn MK et al. (2018) CRISPR-mediated tagging of endogenous proteins with a luminescent peptide. ACS Chem Biol 13(2):467–474. https://doi.org/10.1021/acschembio.7b00549
24. Winter GE et al. (2015) Drug development. Phthalimide conjugation as a strategy for in vivo target protein degradation. Science 348(6241):1376–1381. https://doi.org/10.1126/science.aab1433
25. Burslem GM, Crews CM (2020) Proteolysis-targeting chimeras as therapeutics and tools for biological discovery. Cell 181(1):102–114. https://doi.org/10.1016/j.cell.2019.11.031
26. Miletic N et al. (2025) Workflow for E3 ligase ligand validation for PROTAC development. ACS Chem Biol 20(2):507–521. https://doi.org/10.1021/acschembio.4c00812

Check for updates

Chapter 4

UbiSite Approach: Global Site-Specific Profiling of Canonical and Non-Canonical Ubiquitination in Cells and Tissues

Nicoline Kingo Pedersen, Vyacheslav Akimov and Blagoy Blagoev

Abstract

Ubiquitination is a complex post-translational modification that regulates a wide range of cellular processes through the covalent attachment of ubiquitin to substrate proteins. While canonical ubiquitination occurs on lysine residues, non-canonical modifications on serine, threonine, and the protein N-terminus are increasingly recognized. This chapter describes the UbiSite approach, a refined strategy for site-specific mapping of global ubiquitination using mass spectrometry-based proteomics. The method relies on digestion of the proteome of interest with LysC endopeptidase. LysC digestion leaves a 13 amino acid remnant on the ubiquitinated site of modified peptides, and this remnant is recognized by the monoclonal UbiSite antibody allowing selective enrichment of ubiquitinated peptides. Tens of thousands modification sites are readily identified from cells or tissue extracts in a single UbiSite experiment with the workflow described here. The 13 amino acid remnant is unique to ubiquitin and the UbiSite antibody thereby discriminates ubiquitin from other ubiquitin-like modifiers and enables detection of both canonical and non-canonical ubiquitination. The UbiSite procedure can be automated and combined with label-free quantification as well. UbiSite offers a robust and versatile platform for comprehensive ubiquitinome profiling with enhanced specificity and depth, enabling new insights into the regulatory roles of ubiquitin across diverse biological contexts.

Keywords Mass spectrometry, Ubiquitin, Ubiquitylation, UbiSite, Affinity purification, Canonical ubiquitination, Non-lysine ubiquitination, N-terminal ubiquitination

1 Introduction

1.1 Ubiquitination: A Complex Post-Translational Modification

Ubiquitination is a dynamic and highly regulated post-translational modification (PTM) in which the small protein ubiquitin is covalently attached to substrate proteins. This process is mediated by an enzymatic cascade involving ubiquitin-activating enzymes (E1s), ubiquitin-conjugating enzymes (E2s), and ubiquitin ligases (E3s) [1]. Among these, E3 ligases confer substrate specificity and carry out the final transfer of ubiquitin to target proteins. Ubiquitin can be attached as a single moiety (monoubiquitination) or as polyubiquitin chains,

Maria W. Górna et al. (eds.), *Degrons*, Methods in Molecular Biology, vol. 3069,
https://doi.org/10.1007/978-1-0716-5508-5_4,

which can vary in linkage type, length, and topology, contributing to the functional diversity of this modification [1–3].

Ubiquitination of proteins is involved in virtually all cellular processes [4] and its functional consequences depend on both the site of modification and the linkage type, making it a highly complex and versatile PTM. Ubiquitin is most commonly conjugated to lysine residues, although modifications on the N-terminal amino group of proteins and non-canonical residues such as serine, threonine, and cysteine have also been reported [4–7]. These non-canonical ubiquitination events are less well understood but increasingly recognized for their biological importance [8–11].

1.2 Mapping of Ubiquitination Sites

Global profiling of the endogenous ubiquitinated proteome is technically challenging due to the low abundance and dynamic nature of the modification. Ubiquitinated proteins often exist in substoichiometric quantities relative to their unmodified counterparts, and the modification itself can be rapidly reversed [12, 13]. As a result, sensitive and selective enrichment strategies are essential to isolate ubiquitinated species prior to mass spectrometry (MS)-based identification [8].

One approach is to use protein-level enrichment with, e.g., ubiquitin-binding domains or ubiquitin-specific antibodies [14–16]. While these strategies are effective for capturing ubiquitinated proteins, they are limited by the diversity of ubiquitin linkage types and typically enrich only a subset of the ubiquitinome. Additionally, when the enriched ubiquitinated substrates are proteolytically digested, the majority of the generated peptides will originate from the unmodified regions of the substrate and the attached ubiquitin moieties, which reduces the chance of identifying the actual modification site [17–19].

To achieve site-specific resolution, enrichment at the peptide level has become the preferred strategy. In these strategies, the proteome is enzymatically digested into peptides, and those peptides containing ubiquitin-derived remnants are selectively enriched. The first method developed for this purpose was the use of a diglycine (diGly) antibody, which captures peptides carrying the diGly remnant left on lysines after trypsin digestion [20]. This approach enables high-throughput mapping of canonical ubiquitination sites and has since its development been used in numerous studies to greatly improve our understanding of ubiquitinated proteins [21–25]. However, the diGly antibody has been reported to display sequence-specific biases [23], it cannot capture N-terminal and non-canonical ubiquitination events, and the diGly remnant is not unique to ubiquitin but is shared with other ubiquitin-like modifiers, namely NEDD8 and ISG15 [22, 26]. The Stable Tagged Ubiquitin Exchange (StUbEx)-PLUS approach was developed to overcome some of these limitations [27]. In this approach, the endogenous ubiquitin is knocked down in cells by shRNA and simultaneously replaced by a 6× histidine-tagged version of ubiquitin resistant to the shRNA. The introduced histidine tag is placed internally, in the region between the last lysine of ubiquitin and its C-terminus. When the proteome is digested into peptides using the endopeptidase LysC instead of trypsin, the histidine tag becomes part of the ubiquitin remnant left on the modified sites and can be used for peptide-level enrichment of ubiquitinated peptides. While this allows for discrimination of ubiquitin from other ubiquitin-like modifiers without sequence-specific biases, it requires engineering of the studied system, and it is, therefore, not suited for tissue studies [27].

The UbiSite approach was developed to address the limitations of the current peptide-level strategies [4]. With the UbiSite approach, the proteome is digested into peptides using LysC. This leaves a 13 amino acid remnant on modified sites, which is recognized by the monoclonal UbiSite antibody (Fig. 1A). This longer remnant is unique to ubiquitin, and therefore, allows it to be discriminated from the ISG15 and NEDD8 modifications. Additionally, since the UbiSite antibody is directed towards a region solely within ubiquitin, the capture of ubiquitination events is independent of the ubiquitinated substrate (could even be of non-protein nature) and thus expands the detectable landscape of ubiquitin modifications to include both canonical and potential non-canonical ubiquitination [4, 6, 28].

This chapter provides a comprehensive description of the UbiSite approach (Fig. 1), including: cross-linking of UbiSite antibody to beads (Sect. 3.1), protein extraction from cells or tissues (Sect. 3.2), LysC digestion and peptide clean-up (Sects. 3.3 and 3.4), UbiSite enrichment (Sect. 3.5), sample preparation and LC–MS/MS setup for label-free analysis (Sects. 3.6 and 3.7), and important parameters for the database search (Sect. 3.8).

2 Materials

Prepare buffers and solutions using ultrapure water (18.2 MΩ-cm at 25°C) and use protein low-binding tubes whenever possible.

2.1 Cross-Linking of UbiSite Antibody to Beads

1. Protein G magnetic beads (MagReSyn, MR-PRG002).
2. UbiSite antibody, clone 2G7B8 (Sigma, MABS486).
3. Phosphate-buffered saline (PBS).
4. PBS with 0.1% Triton X-100.
5. 0.2 M sodium borate buffer, pH 9.0: Dissolve 38.14 g of sodium tetraborate decahydrate in 500 mL of water (final volume) and heat to 37°C. Cool down to room temperature (RT), once it is completely dissolved. Adjust pH to 9.0 with 32% of HCl (*see* **Note 1**).

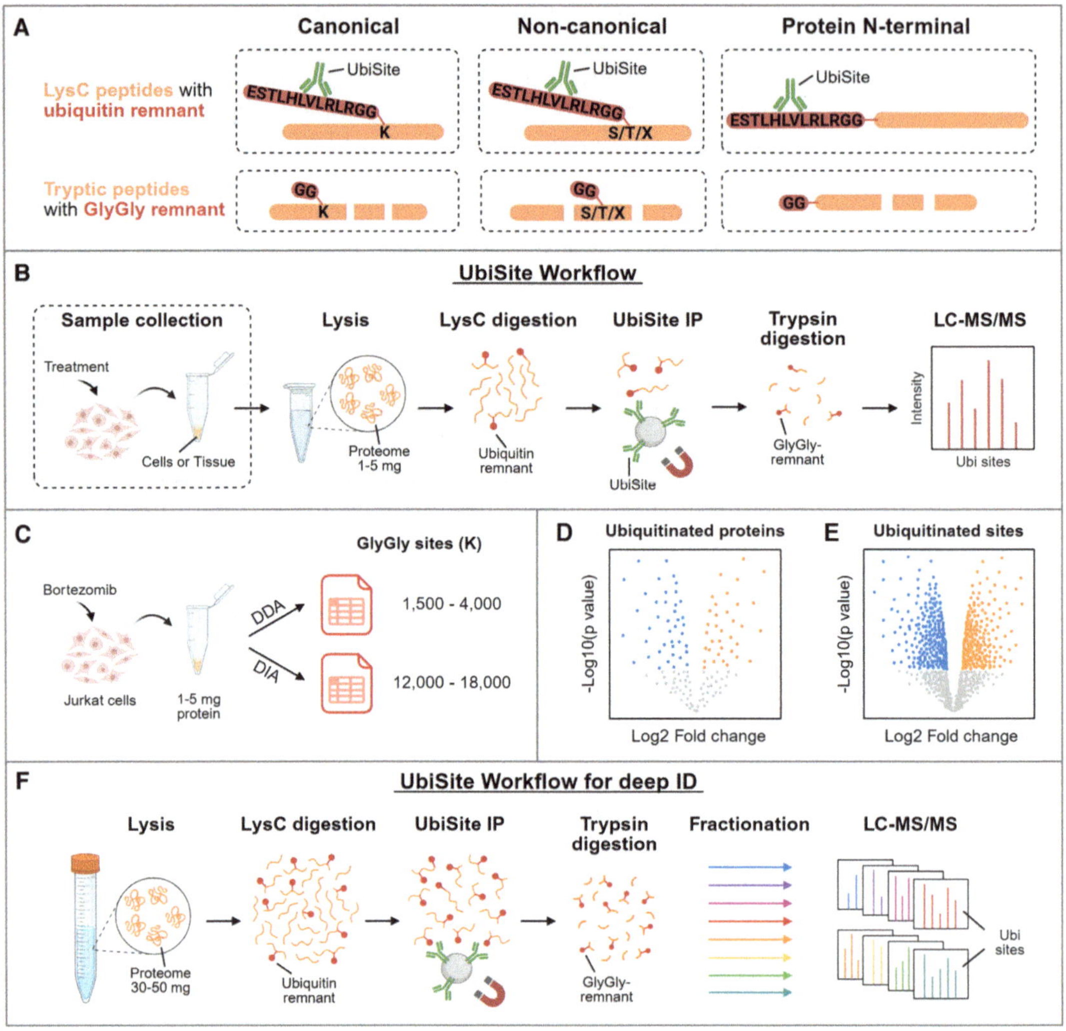

Fig. 1 UbiSite approach for global site-specific profiling of ubiquitination in cells and tissues. (**A**) Illustration of the 13 amino acid ubiquitin remnant that is recognized by the UbiSite antibody after LysC digestion of the proteome. The UbiSite antibody has affinity for the LysC remnant and is suited for enrichment of ubiquitinated canonical lysines (K), protein N-terminus and non-canonical sites including serine (S), threonine (T) or any other amino acid (X). When the LysC peptides are digested with trypsin, a GlyGly signature tag remains on the modified site. (**B**) Workflow of the UbiSite procedure. (**C**) Ubiquitinated lysine sites identified in a single DDA or DIA run on an Exploris 480 from 1–5 mg of protein extracted from Jurkat cells treated with bortezomib. (**D–E**) The UbiSite data allow for downstream analysis of differentially ubiquitinated proteins (**D**) and regulated individual sites (**E**) when comparing different cellular states, treatments, etc. (**F**) Workflow of the UbiSite procedure using increased amount of starting material and off-line fractionation for deeper ubiquitinome coverage. Figure created with BioRender [37]

6. 0.2 M ethanolamine, pH 8.0: Dilute 2.42 mL of 100% ethanolamine stock (16.56 M) in 200 mL of water (final volume). Adjust the pH to 8.0 with HCl (*see* **Note 2**).
7. Dimethyl pimelimidate dihydrochloride (DMP).
8. 10% sodium azide.
9. Magnetic rack.

2.2 Protein Extraction from Cells or Tissue

1. Lysis buffer: 8 M guanidinium-hydrochloride, 50 mM HEPES (pH 8.0).
2. Sonicator (Q500 sonicator from Qsonica or equivalent).
3. BCA protein assay kit and standards.

2.3 Protein Reduction, Alkylation, and LysC Digestion

1. 500 mM tris (2-carboxyethyl)phosphine hydrochloride (TCEP, *see* **Note 3**).
2. 550 mM chloroacetamide (CAA, *see* **Note 4**).
3. 50 mM HEPES, pH 8.5.
4. pH strips.
5. 1 mg/mL LysC (from KPL ApS or equivalent).

2.4 Purification of LysC-Digested Peptides

Use MS-grade solvents.

1. 25% trifluoroacetic acid (TFA).
2. 0.1% TFA.
3. 100% methanol.
4. Solvent B: 60% acetonitrile, 0.1% TFA.
5. C18 cartridge.
6. Syringe.
7. Spectrophotometer.
8. Freeze dryer for lyophilization (ScanVac CoolSafe from LaboGene or equivalent).

2.5 UbiSite Enrichment

1. UbiSite antibodies cross-linked to magnetic beads (*see* **Note 5** and Sect. 2.1 and 3.1).
2. IAP buffer: 50 mM MOPS (pH 7.5), 10 mM sodium phosphate, 50 mM NaCl.
3. Triton X-100.
4. 150 mM NaCl.
5. Low-binding protein tubes.
6. 0.15 % TFA.
7. 0.5 M HEPES, pH 8.0.
8. 0.5 M HEPES, pH 8.5.
9. Trypsin, MS-grade.

10. pH strips.
11. Water bath sonicator.
12. KingFisher Duo Prime with BindIt software (Thermo Scientific) or equivalent.
13. KingFisher Duo Prime plastics for 96 deep-well format (Thermo Scientific) or equivalent.

2.6 Sample Preparation for LC–MS/MS Using StageTips

Use MS-grade solvents

1. 25% TFA.
2. 0.1% TFA.
3. Solvent B: 60% acetonitrile, 0.5% acetic acid.
4. 100% methanol.
5. C18 reversed-phase disks.
6. 200 μL pipette tips.
7. 96-well MS plate.
8. Vacuum centrifuge.
9. pH strips.

2.7 LC–MS/MS Analysis

Use MS-grade solvents

1. Analytical column: 24 cm column packed with 1.9 μm C18 beads.
2. Nano-flow high-performance liquid chromatography (LC) system: EASY-nLC 1000 (Thermo Scientific) or equivalent.
3. Mass Spectrometer (we used Orbitrap Exploris 480 (Thermo Scientific) for the experiments described below).
4. Solvent A: 0.5% acetic acid.
5. Solvent B: 80% acetonitrile, 0.5% acetic acid.

2.8 MS Data Processing

1. MaxQuant version 2.6.4.0 or later (Max Planck Institute of Biochemistry, Germany) for data-dependent acquisition (DDA) data [29, 30].
2. Spectronaut version 18 or later (Biognosys) for data-independent acquisition (DIA) data.

3 Methods

This protocol describes the UbiSite approach starting from collected cell pellets or tissue samples (Fig. 1B). The protocol provided here is optimized for protein starting amounts of 1–5 mg and is readily applicable for label-free quantitation experiments. In the workflow below, we used Jurkat cells treated with proteasomal inhibitors. This usually yields more than 1500 unique sites using DDA and more than 12,000 unique sites with DIA from a single run on an Orbitrap Exploris 480 mass spectrometer (Fig. 1C).

3.1 Cross-Linking of UbiSite Antibody to Beads (see Note 6)

1. Vortex the stock of magnetic Protein G beads and transfer 1 mL of slurry to a 15 mL tube.
2. Wash the beads 2× with 10 mL of PBS.
3. Add 3.6 mg of UbiSite antibody to the beads and adjust the volume to 10 mL with PBS (*see* **Note 7**).
4. Incubate at 4°C for 1.5–3 h with bottom-up rotation at 10–12 RPM.
5. Keep the flow-through and wash the beads 2× with 10 mL of PBS (*see* **Note 8**).
6. Wash the beads 2× with 10 mL of 0.2 M sodium borate buffer.
7. Add fresh 12 mL of 0.2 M sodium borate buffer to the beads.
8. Weigh ~70–80 mg of DMP (*see* **Note 9**) and add it directly to the bead-buffer solution. Mix immediately.
9. Incubate for 45 min at RT with rotation at 10 RPM.
10. Remove the DMP solution and wash the beads once with 0.2 M sodium borate buffer. Discard the buffer.
11. Add 12 mL of 0.2 M ethanolamine for quenching. Incubate for 1.5 h at RT with rotation at 10 RPM (*see* **Note 10**).
12. Wash the beads 3× with 10 mL of PBS containing 0.1% Triton X-100.
13. Resuspend the beads in PBS containing 0.1% Triton X-100 to a final slurry volume of 1 mL (*see* **Note 8**).
14. Add sodium azide to a final concentration of 0.05% and store at 4°C.

3.2 Protein Extraction from Cells or Tissue (see Note 11)

1. Resuspend samples in RT lysis buffer at a 1:3 ratio (sample:buffer).
2. Sonicate for 30 s at 20% amplitude. Repeat if samples remain viscous.
3. Centrifuge 16,000 RCF for 15 min at RT. Transfer the supernatant to new tubes.
4. Measure protein concentrations using a BCA assay.
5. Use 1–5 mg of protein (*see* **Note 12**).

3.3 Protein Reduction, Alkylation, and LysC Digestion

1. Reduce and alkylate proteins by incubating samples with a final concentration of 5 mM TCEP and 11 mM CAA (*see* **Note 4**) for 30 min at RT protected from light (*see* **Note 13**).
2. Dilute lysates to 2 M guanidinium-hydrochloride with 50 mM HEPES, pH 8.5. Confirm that pH is 8.5 with a pH strip.

3. Add LysC at a 1:100 ratio (LysC:protein, e.g., 10 μg of LysC per 1 mg of protein). Incubate overnight (~16 h) at RT (*see* **Note 14**).

3.4 Purification of LysC-Digested Peptides

Use a C18 cartridge with appropriate binding capacity (*see* **Note 15**). Take up the specified volume of solution in a syringe and attach it to the cartridge. Apply pressure to push the solutions through with 1–2 drops/s. Avoid drying the cartridge during equilibration, washing, and sample loading.

1. Wash one cartridge per sample with 4 mL of 100% methanol.
2. Wash with 4 mL of Solvent B.
3. Wash and equilibrate 2× with 12 mL of 0.1% TFA.
4. Acidify the LysC digested samples by adding TFA to a final concentration of 1% and confirm that the pH~2.0 with a pH strip.
5. Centrifuge samples at 3000 RCF for 5–10 min at RT.
6. Collect supernatant using a syringe and needle, avoiding the pellet.
7. Load the sample onto the equilibrated cartridge by applying pressure to push the sample through with 1 drop/s.
8. Wash the cartridge 2× with 12 mL of 0.1% TFA.
9. Elute the peptides with 2 × 2 mL of Solvent B into a 15 mL tube. Make sure to collect all the elution by flushing air through the cartridge. Combine the elutions and vortex gently.
10. Measure the peptide concentration in the elution with a UV spectrophotometer.
11. Prepare samples for lyophilization of the peptides: Freeze the samples at −80°C for 8–16 h. Freeze the tubes tilted to maximize the free surface of the samples, which will speed up the subsequent lyophilization process.
12. When the samples are frozen, make a small hole in the tube caps and lyophilize for 12–24 h (*see* **Note 16**).

3.5 UbiSite Enrichment

1. Dissolve lyophilized peptides in IAP buffer to have a peptide concentration of 2–5 mg/mL. The peptides should be dissolved in a maximum of 1 mL, which is the capacity of the wells in the KingFisher plates used for automated enrichment.
2. Add Triton X-100 to a final concentration of 0.2% and vortex.
3. Let peptides dissolve for 5 min. If the solution is not clear but cloudy, sonicate samples for 5 min in a water bath at RT.

4. Confirm that the pH is 7.5–8.0 with a pH strip. If the samples are acidic, add HEPES, pH 8.0, to a final concentration of 10 mM.
5. Centrifuge samples at 16,000 RCF for 10 min at 4°C. Proceed with automated UbiSite enrichment using the KingFisher Duo Prime (*see* **Note 17** for a description of manual handling).
6. Prepare KingFisher plates and elution strips according to the schematic overview in Table 1.
7. Use 20 μL of UbiSite magnetic bead slurry per 3 mg peptide (*see* **Note 18**).
8. Run the KingFisher program as described in Table 2 (*see* **Note 19**).

Table 1
Overview of the content of the KingFisher (KF) plates and elution strips used for automated UbiSite enrichment, subsequent washes, and elution with the KingFisher Duo Prime

Plate/strip name	Well	Name of well	Content information	Volume
KF plate 1*	A	LysC peptides	LysC peptides in IAP + 0.2% Triton X-100	1 mL
	B	UbiSite beads	UbiSite beads in IAP + 0.2% Triton X-100	1 mL
	C	IAP wash 1	IAP + 0.2% Triton X-100	1 mL
	D	IAP wash 2	IAP + 0.2% Triton X-100	1 mL
	E	IAP wash 3	IAP + 0.2% Triton X-100	1 mL
	F	IAP wash 4	IAP	1 mL
	G	Store used beads	IAP	500 μL
	H	Comb 1	Place comb 1	–
KF plate 2*	A	IAP wash 5	IAP	1 mL
	B	IAP wash 6	IAP	1 mL
	C	IAP wash 7	IAP	1 mL
	D	IAP wash 8	IAP	500 μL
	E	IAP wash 9	IAP	500 μL
	F	IAP wash 10	IAP	500 μL
	G	NaCl wash 1	150 mM NaCl	500 μL
	H	Comb 2	Place comb 2	–
Elution strip 1**	A	Elution 1	0.15 % TFA	60 μL
Elution strip 2**	A	Elution 2	0.15 % TFA	60 μL

*Use cold solutions, **Use solutions at RT

Table 2
Suggested settings for automated UbiSite enrichment using the KingFisher Duo Prime. The program takes approximately 3.5 h

Step	Protocol step	Name of well	KingFisher settings
1	Pickup Comb 1	KF plate 1 (H)	
2	Collect UbiSite beads	KF plate 1 (B)	Collect count: 5, Collect time: 30 s
3	Pulldown of LysC peptides	KF plate 1 (A)	Release beads: 10 s, speed = slow Mix time: 2 h, speed = slow, temperature = 10°C Collect beads: Count: 5, Collect time: 30 s
4-9	IAP wash 1–6	KF plate 1 (C-F) KF plate 2 (A-B)	Release beads: 10 s, speed = slow Mix time: 1 min, speed = slow Collect beads: Count: 5, Collect time: 30 s
10	IAP wash 7	KF plate 2 (C)	Release beads: 10 s, speed = slow Mix time: 1 min, speed = slow (do not collect beads)
11	Leave Comb 1	KF plate 1 (B)	
12	Pickup Comb 2	KF plate 2 (H)	
13	Collect UbiSite beads	KF plate 2 (C)	Collect count: 5, Collect time: 30 s
14–16	IAP wash 8–10	KF plate 2 (D-F)	Release beads: 10 s, speed = slow Mix time: 1 min, speed = slow Collect beads: Count: 5, Collect time: 30 s
17	NaCl wash 1	KF plate 2 (G)	Release beads: 10 s, speed = slow Mix time: 30 s, speed = slow Collect beads: Count: 5, Collect time: 30 s
18	Elution 1	Elution strip 1 (A)	Release beadsMix time: 7.5 min, speed = FastCollect beads
19	Elution 2	Elution strip 2 (A)	Release beads Mix time: 7.5 min, speed = Fast Collect beads
20	Leave Comb2 and UbiSite beads	KF plate 1 (G)	

9. *Optional*: The eluted LysC-digested peptides can be prepared directly for LC–MS/MS (proceed to Sect. 3.6); however, we recommend to further digest the enriched peptides with trypsin, as described below (*see* **Note 20**).
10. Neutralize eluted peptides by adding HEPES, pH 8.5, to a final concentration of 100 mM. Confirm pH 8.5 with a pH strip. Use as little sample as possible for the pH test (0.5 μL).
11. Add 75 ng of trypsin per 3 mg of the peptide starting amount (measured in Sect. 3.4, Step 10).

12. Mix by flicking the tube, spin down, and incubate overnight at 37°C.

3.6 Sample Preparation for LC–MS/MS Using StageTips

1. Acidify the samples by adding TFA to a final concentration of 0.5%. Confirm pH ~2.0.
2. Prepare one StageTip per sample by packing 2× C18 disks into a 200 μL pipette tip [31] (*see* **Note 21**).
3. Wash the StageTip 1× with 10 μL of 100% methanol.
4. Wash 1× with 30 μL of Solvent B.
5. Wash and equilibrate 2× with 30 μL of 0.1% TFA.
6. Load sample.
7. Wash 2× with 30 μL of 0.1% TFA.
8. Elute peptides 2× with 20 μL of Solvent B into a 0.5 mL tube.
9. Evaporate solvent from the elutions in a vacuum centrifuge at RT until 0.5–1 μL of the elution is left.
10. Resuspend peptides in an appropriate volume of 0.1% TFA and transfer the samples into a 96-well MS plate. The peptides are now ready for LC–MS/MS.

3.7 LC–MS/MS Analysis

Historically, DDA has been the standard for PTM analysis, offering high-quality spectra by selectively fragmenting the most abundant precursor ions. However, DDA is inherently stochastic, often resulting in missing values across replicates [32]. With the newer generations of mass spectrometers, DIA performs equally well or better than DDA for ubiquitin analysis [28]. DIA fragments all ions within defined mass-to-charge (m/z) windows, providing more

Table 3
Suggested gradient for nano-LC–MS/MS analysis of tryptic ubiquitinated peptides (DDA and DIA)

Time (min)	Flow rate (nL/min)	% Solvent B[a]
0	250	7
3	250	7
10	250	12
96	250	35
111	250	45
112	250	95
117	250	95
118	250	7
120	250	7

[a]Solvent B = 80% Acetonitrile, 0.5% acetic acid

comprehensive and reproducible coverage of the ubiquitin-modified proteome. Combining the UbiSite approach with label-free DIA allows for deep ubiquitinome profiling from as little as 1–5 mg of protein starting material, without compromising data quality (*see* **Note 12**).

The LC–MS/MS setup used in this protocol includes an Orbitrap Exploris 480 mass spectrometer coupled to an EASY-nLC 1000 LC system. Acquisition parameters are detailed in Table 3 (LC settings) and Table 4 (MS settings).

3.8 MS Data Processing

Several advanced software tools are compatible with the identification and quantification of ubiquitin-modified peptides. In this protocol, MaxQuant is used for DDA data and Spectronaut for DIA data.

3.8.1 Peptide and Protein Identification from DDA Data Using MaxQuant

1. Open MaxQuant version 2.6.4.0 or later versions [29, 30].
2. Load raw DDA files and set experiment.
3. In the *Group-specific parameter* module, configure the following parameters:

Table 4
Suggested MS and MS/MS acquisition parameters for the analysis of tryptic ubiquitin peptides using DDA and DIA on an Orbitrap Exploris 480 mass spectrometer

Parameter	DDA	DIA
Ionization mode	Positive	Positive
MS1 orbitrap resolution	60 K	120 K
MS1 scan range (m/z)	350–1500	350–1400
MS1 normalized AGC target (%)	300	300
MS1 maximum Injection time (ms)	25	45
MS2 mass range	–	361–1033
MS2 isolation window (m/z)	1.2	12
MS2 window overlap (m/z)	–	1
MS2 scan events	–	56
MS2 orbitrap resolution	30 K	30 K
MS2 normalized AGC target (%)	200	1000
MS2 maximum Injection time (ms)	54	54
HCD collision energy (normalized, %)	30	28
Intensity threshold	$1.0e^4$	–
Included charge states	2–6	–
Dynamic exclusion	60 s with a ±10 ppm window	–
DDA method	Top 12 number of scans	–

   a. Go to *Digestion* and choose Trypsin/P as enzyme.
   b. Go to *Modifications* and add the Acetyl (Protein N-term), Oxidation (M), and GlyGly (K) [+114.04 Da] as variable modifications (*see* **Note 22**).

4. In the *Global parameters* module, go to *Sequences* and load an appropriate protein database.
5. Start the search.
6. When the search is completed, download the.txt files for identification and quantitation of unique GlyGly-modified peptides.
7. Proceed with downstream data analysis (*see* **Note 23**).

3.8.2 Peptide and Protein Identification from DIA Data Using Spectronaut

1. Open Spectronaut version 18 or later versions.
2. Set up a DirectDIA™ analysis (library-free workflow) [33].
3. Load the DIA raw files and select the appropriate protein database.
4. Select the search and extraction settings schema for direct-DIA™ analysis using the BGS Factory Settings (default) as a template.
5. In the *Pulsar search* module, configure the following parameters:
   a. Go to *Peptides* and choose Trypsin/P as enzyme.
   b. Go to *Modifications* and add the Acetyl (Protein N-term), Oxidation (M), and GlyGly (K) [+114.04 Da] as variable modifications (*see* **Note 22**).
6. In the *DIA analysis* module, configure the following parameters:
   a. Go to *Quantification* and enable the *Cross-Run Normalization*; set Normalization Filter Type to None (*see* **Note 24**).
   b. Go to *PTM Workflow* and enable *PTM Localization*, with a probability cutoff of 0.75; enable PTM Analysis, Hierarchical Clustering, and Multiplicity.
7. Specify experimental conditions.
8. Start the search.
9. When the search is completed, generate a report containing columns for identification and quantitation of unique GlyGly-modified peptides.
10. Proceed with downstream data analysis (*see* **Notes 23 and 25**, and Fig. 1D–E).

4 Notes

1. The sodium borate buffer may show slight crystallization. Make sure it is completely dissolved before use.
2. Adjust the pH to 8.0 slowly with HCl. Small additions of acid can cause large pH shifts.
3. TCEP is purchased as a 0.5 M aqueous solution at pH 7.0. It is sensitive to oxidation, make sure to use fresh aliquots for optimal reducing efficiency. Store unopened vials at 4°C protected from light. After opening, aliquot and store at −20°C (long-term) or 4°C (short-term). Dithiothreitol (DTT) may be used as an alternative reducing agent (*see* **Note 13**).
4. CAA is light-sensitive, and stocks should be stored at −20°C protected from light. Iodoacetamide (IAA) is another commonly used alkylation reagent; however, it is not recommended for the UbiSite procedure or for ubiquitination analysis overall. The reactivity of IAA is much greater than CAA, which drastically increases the risk of side reactions between IAA and unmodified lysines, forming a 2-acetamidoacetamide modification with a mass identical to the diGly remnant [34].
5. Various types of Protein G beads are suitable for enrichment. We have evaluated multiple agarose-based and magnetic Protein G bead formulations. Optimal bead performance is characterized by a high antibody binding capacity and minimal non-specific binding. Based on our observations, agarose beads generally exhibit superior antibody binding efficiency. However, magnetic beads are compatible with automated workflows, while performing similarly to agarose beads in terms of specificity and yield.
6. The beads can be prepared in advance and stored at 4°C with sodium azide for at least 1 year. Do not freeze. The cross-linking procedure is the same for other types of magnetic or agarose beads. We recommend to cross-link the UbiSite antibody to beads to prevent coelution of antibodies with the enriched peptides. Co-elution of the antibody will result in high contaminating background during MS analysis, hindering the identification of ubiquitinated peptides.
7. Assess the antibody binding capacity of the selected beads before use. Add 5%–10% more antibody than the estimated binding capacity to ensure saturation of the beads. It will help maximize enrichment efficiency and minimize variability in enrichment performance.
8. To verify cross-linking efficiency, perform Western blotting using 30 μL of the flow-through from Sect. 3.1, Step 5

together with an aliquot of the cross-linked beads corresponding to 3–5 μg of UbiSite antibody (from Sect. 3.1, Step 13). Use a mouse secondary antibody for probing the Western blot. With successful cross-linking, only the light antibody chain should be observed. Both the light and heavy antibody chains may appear in the flow-through and can be used to estimate if the beads were completely saturated with antibody.

9. Up to 95 mg of DMP is acceptable. The compound may be sticky during weighing, and a slight excess of DMP does not compromise the efficiency of the procedure.

10. Beads may clump during quenching. Make sure to resuspend thoroughly before proceeding.

11. Samples can be stored at −20°C in lysis buffer, but it is recommended to proceed to lyophilization of LysC peptides (Sects. 3.2–3.4), which is the best stage in the protocol for storing the samples.

12. It is crucial to measure the protein concentration to estimate the correct amount of LysC to be used for digestion. For the Orbitrap Exploris 480 instrument, 1–5 mg starting protein material is usually sufficient for DIA-based profiling of ubiquitinated sites; however, the depth will depend on sample types, treatments, etc. In our hands, 1–5 mg of protein extracted from cells treated with a proteasome inhibitor usually provide more than 12,000 identified sites from a single run. Deeper identification can be achieved with faster and more sensitive instruments like the Orbitrap Astral mass spectrometer (Thermo Scientific), which to our experience, results in >2 times more ubiquitinated sites in a single run. Alternatively, a deeper ubiquitinome coverage can be achieved by increasing the starting material, combined with offline fractionation of the sample, e.g., by high pH reversed phase chromatography, and then each fraction can be analyzed separately by LC–MS/MS (Fig. 1F) [4, 35].

13. If DTT is used as a reducing reagent instead of TCEP, incubate with a final concentration of 2 mM DTT for 30 min at RT before alkylation with CAA for 30 min at RT [4].

14. After LysC digestion, samples may appear milky due to undigested bigger peptides, chromatin leftover, lipids, or cell debris. This is normal—proceed to the next step in the procedure.

15. Use a C18 cartridge with appropriate binding capacity for the peptide amount. The volumes specified in the protocol are for the use of one cartridge per sample (3 mg peptides). Volumes can be scaled up if more cartridges are required per sample.

16. To our experience, lyophilization is critical for the enrichment efficiency and should not be substituted with other means of reducing the volume or drying. Correctly lyophilized peptides should appear as powder or flakes, not solid traces.
17. Manual enrichment protocol using a magnetic rack as an alternative to the KingFisher automation. If agarose beads (not magnetic) are used instead, the beads are pelleted with a table microcentrifuge:
    - Use 20 μL of UbiSite bead slurry per 3 mg of peptides (*see* **Note 18**).
    - Wash beads 1× with cold IAP buffer containing 0.2% Triton X-100.
    - Incubate beads with sample for 1.5 h at 4°C with rotation at 10 RPM (*see* **Note 19**).
    - Wash beads 3× with 1 mL of cold IAP buffer containing 0.2% Triton X-100 for 3 min with rotation at 4°C.
    - Wash beads 3× with 1 mL of cold IAP buffer by gently flipping the tube. Allow sufficient time for the beads to reach the magnet. It usually takes longer when there is no detergent in the buffer, as the beads will stick more to the plastic. It can be beneficial to flip the tube while the samples are on the magnet.
    - Resuspend beads in 1 mL of cold IAP buffer and transfer to a new protein low-binding 1.5 mL tube.
    - Wash beads 3× with 1 mL of cold IAP buffer.
    - Wash 1× with 1 mL of 150 mM NaCl.
    - Elute peptides by adding 30 μL of 0.15% TFA and incubate for 5 min at RT. Mix beads by flicking the tube every minute. Transfer elution to a new protein low-binding tube.
    - Repeat the elution step two more times and combine with the first elution.
    - Proceed to Sect. 3.5, Step 9.
18. In this protocol, 20 μL of UbiSite bead slurry corresponds to 75 μg of UbiSite antibody and is used per 3 mg of peptide. If other beads are used, adjust the volume based on the antibody-binding capacity of the beads to have 75–100 μg of UbiSite antibody per 3 mg of peptide. Scale up or down accordingly.
19. Do not incubate the UbiSite beads with LysC peptides overnight, as the prolonged incubation increases non-specific binding of unmodified peptides. If you use agarose beads, incubate for 3 h.

20. After elution, the enriched LysC peptides are further digested with trypsin to enhance MS compatibility and identification [4, 36]. The LysC peptides can indeed be analyzed directly; however, these peptides are generally longer and carry a longer ubiquitin remnant, resulting in fewer identified ubiquitinated sites. Since the selective enrichment of ubiquitinated peptides is already completed at this stage, we recommend continuing with trypsin digestion as described in detail in the step-by-step procedure. This dual digestion improves peptide fragmentation and identification efficiency, facilitating deeper and more accurate profiling of the ubiquitin-modified proteome [4, 36].
21. We make StageTips using a syringe-like tool to cut C18 disks with ~1 mm diameter, and two disks are inserted into a p200 tip [31]. StageTips are also commercially available (e.g., CDS Empore).
22. To add a new modification, open the *Configuration* module in MaxQuant or the *Databases* module in Spectronaut, select *Modification* and add new. Give the modification an appropriate name (GlyGly (K)) and add the composition of the GlyGly remnant (C4H6N2O2) with a mass shift of 114.04 Da. Specify the position to be not C-term and choose K as the modification site, or add other sites if non-canonical modifications are of interest, e.g., S or T. Save the modification and use it in the search. For N-terminal modifications, specify the position to be the protein N-term and choose all amino acids as site.
23. It is good practice to do quality control of the data including a check of the enrichment efficiency. Because the UbiSite enrichment is performed on LysC peptides and subsequently digested with trypsin, an optimal enrichment efficiency will result in 30%–40% of tryptic peptides containing a GlyGly modification, while the remaining peptides are flanking peptides to the modified site [4]. Additionally, ensure that only unique sites are included in the subsequent analysis by removing redundancies.
24. For the UbiSite approach, it is not advised to change the *Normalization Filter Type* to *Modification Type Filter*. This filter type is normally recommended for samples with high PTM enrichment efficiency. UbiSite enrichment usually yields 30%–40% of tryptic peptides containing a GlyGly modification, while the remaining peptides are flanking peptides to the modified site (*see* **Note 23**). Both the modified peptides and the flanking peptides originate from the enriched sample and can, therefore, be included in the normalization.

25. The generated data are well suited for differential ubiquitination analysis—both at protein and individual site level—when comparing, e.g., different cellular states, drug responses, treatments, etc. (Fig. 1D–E).

Acknowledgments

The work was funded partially from the Danish National Research Foundation (grant no. DNRF141), Independent Research Fund Denmark (grant no. 1026-00013B), PRO-MS: Danish National Mass Spectrometry Platform for Functional Proteomics (grant no. 5072–00007B) and INTEGRA (Novo Nordisk Foundation, grant no. NNF20OC0061575).

References

1. Komander D, Rape M (2012) The ubiquitin code. Annu Rev Biochem 81:203–229. https://doi.org/10.1146/annurev-biochem-060310-170328
2. Phu L, Izrael-Tomasevic A, Matsumoto ML, Bustos D, Dynek JN, Fedorova AV, Bakalarski CE, Arnott D, Deshayes K, Dixit VM, Kelley RF, Vucic D, Kirkpatrick DS (2011) Improved quantitative mass spectrometry methods for characterizing complex ubiquitin signals. Mol Cell Proteomics 10(5):M110.003756. https://doi.org/10.1074/mcp.M110.003756
3. Clague MJ, Heride C, Urbé S (2015) The demographics of the ubiquitin system. Trends Cell Biol 25(7):417–426. https://doi.org/10.1016/j.tcb.2015.03.002
4. Akimov V, Barrio-Hernandez I, Hansen SVF, Hallenborg P, Pedersen AK, Bekker-Jensen DB, Puglia M, Christensen SDK, Vanselow JT, Nielsen MM, Kratchmarova I, Kelstrup CD, Olsen JV, Blagoev B (2018) UbiSite approach for comprehensive mapping of lysine and N-terminal ubiquitination sites. Nat Struct Mol Biol 25(7):631–640. https://doi.org/10.1038/s41594-018-0084-y
5. Bhogaraju S, Kalayil S, Liu Y, Bonn F, Colby T, Matic I, Dikic I (2016) Phosphoribosylation of ubiquitin promotes serine ubiquitination and impairs conventional ubiquitination. Cell 167(6):1636–1649. https://doi.org/10.1016/j.cell.2016.11.019 e1613
6. McCrory EH, Akimov V, Cohen P, Blagoev B (2022) Identification of ester-linked ubiquitylation sites during TLR7 signalling increases the number of inter-ubiquitin linkages from 8 to 12. Biochem J 479(23):2419–2431. https://doi.org/10.1042/bcj20220510
7. Pao KC, Wood NT, Knebel A, Rafie K, Stanley M, Mabbitt PD, Sundaramoorthy R, Hofmann K, van Aalten DMF, Virdee S (2018) Activity-based E3 ligase profiling uncovers an E3 ligase with esterification activity. Nature 556(7701):381–385. https://doi.org/10.1038/s41586-018-0026-1
8. van Overbeek NK, Aguirre T, van der Heden van Noort GJ, Blagoev B, Vertegaal ACO (2023) Deciphering non-canonical ubiquitin signaling: biology and methodology. Front Mol Biosci 10:1332872. https://doi.org/10.3389/fmolb.2023.1332872
9. Kelsall IR (2022) Non-lysine ubiquitylation: doing things differently. Front Mol Biosci 9:1008175. https://doi.org/10.3389/fmolb.2022.1008175
10. Squair DR, Virdee S (2022) A new dawn beyond lysine ubiquitination. Nat Chem Biol 18(8):802–811. https://doi.org/10.1038/s41589-022-01088-2
11. Akizuki Y, Kaypee S, Ohtake F, Ikeda F (2024) The emerging roles of non-canonical ubiquitination in proteostasis and beyond. J Cell Biol 223(5):e202311171. https://doi.org/10.1083/jcb.202311171
12. Ordureau A, Münch C, Harper JW (2015) Quantifying ubiquitin signaling. Mol Cell 58(4):660–676. https://doi.org/10.1016/j.molcel.2015.02.020
13. Prus G, Satpathy S, Weinert BT, Narita T, Choudhary C (2024) Global, site-resolved analysis of ubiquitylation occupancy and turnover rate reveals systems properties. Cell 187(11):2875–2892. https://doi.org/10.1016/j.cell.2024.03.024
14. Akimov V, Fehling-Kaschek M, Barrio-Hernandez I, Puglia M, Bunkenborg J, Nielsen MM, Timmer J, Dengjel J, Blagoev B (2021) Magnitude of ubiquitination determines the fate of epidermal growth factor receptor upon

ligand stimulation. J Mol Biol 433(21):167240. https://doi.org/10.1016/j.jmb.2021.167240
15. Hjerpe R, Aillet F, Lopitz-Otsoa F, Lang V, England P, Rodriguez MS (2009) Efficient protection and isolation of ubiquitylated proteins using tandem ubiquitin-binding entities. EMBO Rep 10(11):1250–1258. https://doi.org/10.1038/embor.2009.192
16. Newton K, Matsumoto ML, Wertz IE, Kirkpatrick DS, Lill JR, Tan J, Dugger D, Gordon N, Sidhu SS, Fellouse FA, Komuves L, French DM, Ferrando RE, Lam C, Compaan D, Yu C, Bosanac I, Hymowitz SG, Kelley RF, Dixit VM (2008) Ubiquitin chain editing revealed by polyubiquitin linkage-specific antibodies. Cell 134(4):668–678. https://doi.org/10.1016/j.cell.2008.07.039
17. Hitchcock AL, Auld K, Gygi SP, Silver PA (2003) A subset of membrane-associated proteins is ubiquitinated in response to mutations in the endoplasmic reticulum degradation machinery. Proc Natl Acad Sci U S A 100(22):12735–12740. https://doi.org/10.1073/pnas.2135500100
18. Danielsen JM, Sylvestersen KB, Bekker-Jensen S, Szklarczyk D, Poulsen JW, Horn H, Jensen LJ, Mailand N, Nielsen ML (2011) Mass spectrometric analysis of lysine ubiquitylation reveals promiscuity at site level. Mol Cell Proteomics 10(3):M110.003590. https://doi.org/10.1074/mcp.M110.003590
19. Akimov V, Henningsen J, Hallenborg P, Rigbolt KT, Jensen SS, Nielsen MM, Kratchmarova I, Blagoev B (2014) StUbEx: stable tagged ubiquitin exchange system for the global investigation of cellular ubiquitination. J Proteome Res 13(9):4192–4204. https://doi.org/10.1021/pr500549h
20. Xu G, Paige JS, Jaffrey SR (2010) Global analysis of lysine ubiquitination by ubiquitin remnant immunoaffinity profiling. Nat Biotechnol 28(8):868–873. https://doi.org/10.1038/nbt.1654
21. Wagner SA, Beli P, Weinert BT, Nielsen ML, Cox J, Mann M, Choudhary C (2011) A proteome-wide, quantitative survey of in vivo ubiquitylation sites reveals widespread regulatory roles. Mol Cell Proteomics 10(10):M111.013284. https://doi.org/10.1074/mcp.M111.013284
22. Kim W, Bennett EJ, Huttlin EL, Guo A, Li J, Possemato A, Sowa ME, Rad R, Rush J, Comb MJ, Harper JW, Gygi SP (2011) Systematic and quantitative assessment of the ubiquitin-modified proteome. Mol Cell 44(2):325–340. https://doi.org/10.1016/j.molcel.2011.08.025
23. Wagner SA, Beli P, Weinert BT, Schölz C, Kelstrup CD, Young C, Nielsen ML, Olsen JV, Brakebusch C, Choudhary C (2012) Proteomic analyses reveal divergent ubiquitylation site patterns in murine tissues. Mol Cell Proteomics 11(12):1578–1585. https://doi.org/10.1074/mcp.M112.017905
24. Fulzele A, Bennett EJ (2018) Ubiquitin diGLY proteomics as an approach to identify and quantify the ubiquitin-modified proteome. Methods Mol Biol 1844:363–384. https://doi.org/10.1007/978-1-4939-8706-1_23
25. Udeshi ND, Mertins P, Svinkina T, Carr SA (2013) Large-scale identification of ubiquitination sites by mass spectrometry. Nat Protoc 8(10):1950–1960. https://doi.org/10.1038/nprot.2013.120
26. Na CH, Jones DR, Yang Y, Wang X, Xu Y, Peng J (2012) Synaptic protein ubiquitination in rat brain revealed by antibody-based ubiquitome analysis. J Proteome Res 11(9):4722–4732. https://doi.org/10.1021/pr300536k
27. Akimov V, Olsen LCB, Hansen SVF, Barrio-Hernandez I, Puglia M, Jensen SS, Solov'yov IA, Kratchmarova I, Blagoev B (2018) StUbEx PLUS-A modified stable tagged ubiquitin exchange system for peptide level purification and in-depth mapping of ubiquitination sites. J Proteome Res 17(1):296–304. https://doi.org/10.1021/acs.jproteome.7b00566
28. Trulsson F, Akimov V, Robu M, van Overbeek N, Berrocal DAP, Shah RG, Cox J, Shah GM, Blagoev B, Vertegaal ACO (2022) Deubiquitinating enzymes and the proteasome regulate preferential sets of ubiquitin substrates. Nat Commun 13(1):2736. https://doi.org/10.1038/s41467-022-30376-7
29. Cox J, Neuhauser N, Michalski A, Scheltema RA, Olsen JV, Mann M (2011) Andromeda: a peptide search engine integrated into the MaxQuant environment. J Proteome Res 10(4):1794–1805. https://doi.org/10.1021/pr101065j
30. Tyanova S, Temu T, Cox J (2016) The MaxQuant computational platform for mass spectrometry-based shotgun proteomics. Nat Protoc 11(12):2301–2319. https://doi.org/10.1038/nprot.2016.136
31. Ishihama Y, Rappsilber J, Mann M (2006) Modular stop and go extraction tips with stacked disks for parallel and multidimensional Peptide fractionation in proteomics. J Proteome Res 5(4):988–994. https://doi.org/10.1021/pr050385q
32. Guo T, Steen JA, Mann M (2025) Mass-spectrometry-based proteomics: from single cells to clinical applications. Nature 638(8052):901–911. https://doi.org/10.1038/s41586-025-08584-0
33. Martinez-Val A, Bekker-Jensen DB, Hogrebe A, Olsen JV (2021) Data Processing and Analysis for DIA-Based Phosphoproteomics Using Spectronaut.

Methods Mol Biol 2361:95–107. https://doi.org/10.1007/978-1-0716-1641-3_6
34. Nielsen ML, Vermeulen M, Bonaldi T, Cox J, Moroder L, Mann M (2008) Iodoacetamide-induced artifact mimics ubiquitination in mass spectrometry. Nat Methods 5(6):459–460. https://doi.org/10.1038/nmeth0608-459
35. Batth TS, Francavilla C, Olsen JV (2014) Off-line high-pH reversed-phase fractionation for in-depth phosphoproteomics. J Proteome Res 13(12):6176–6186. https://doi.org/10.1021/pr500893m
36. Hendriks IA, Akimov V, Blagoev B, Nielsen ML (2021) MaxQuant.Live enables enhanced selectivity and identification of peptides modified by endogenous SUMO and Ubiquitin. J Proteome Res 20(4):2042–2055. https://doi.org/10.1021/acs.jproteome.0c00892
37. Blagoev B (2025) https://BioRender.com/rz04wul. Created in BioRender

Part II

Synthesis and In Vitro Studies of Degrons

Chapter 5

Chemical and Enzymatic Approaches to C-Terminal Cyclic Imide Formation

Zhenguang Zhao, Ethan Yang Feng, Wenqing Xu and Christina M. Woo

Abstract

The C-terminal cyclic imide modification is recognized as a degron by the E3 ligase adapter cereblon (CRBN). Reliable methods to generate and measure this modification on CRBN substrates are essential for studying its biological function. In this chapter, we describe two complementary approaches for producing C-terminal cyclic imide-modified peptides and proteins: sortase-mediated transpeptidation and enzymatic conversion by protein carboxymethyltransferase (PCMT1). Additionally, we present analytical strategies, including a time-resolved Förster resonance energy transfer (TR-FRET) assay for measuring binding of substrates to CRBN in vitro and mass spectrometry methods for confirming cyclic imide formation. Together, these protocols provide practical methods for investigating the chemistry and biology of C-terminal cyclic imides.

Keywords C-terminal cyclic imide, Cereblon, Targeted protein degradation, PCMT1, Sortase, TR-FRET, Mass spectrometry

1 Introduction

Post-translational modifications (PTMs) expand the chemical diversity of proteins and play crucial roles in regulating their stability, localization, and function. Among these, the C-terminal cyclic imide, derived from the intramolecular cyclization of asparagine or glutamine residues (denoted cN or cQ, respectively), is a PTM that is now gaining attention due to its association as a degradation signal, or "degron", recognized by the E3 ligase substrate adapter cereblon (CRBN) [1]. C-Terminal cyclic imides, in particular C-terminal cN (*see* **Note 1**), can form spontaneously under physiological conditions [2] or enzymatically through the activity of protein carboxymethyltransferase (PCMT1) [3]. Once present, CRBN recognizes the C-terminal cyclic imide in both endogenous and engineered substrates to

The authors "Zhenguang Zhao, Ethan Yang Feng, and Wenqing Xu" contributed equally.

Maria W. Górna et al. (eds.), *Degrons*, Methods in Molecular Biology, vol. 3069,
https://doi.org/10.1007/978-1-0716-5508-5_5,

promote their proteasomal degradation. These findings have sparked a need for methods to generate and characterize proteins bearing this modification.

Beyond their function as a degron on proteins, C-terminal cyclic imides have applications in targeted protein degradation. Therapeutics such as thalidomide and its analogues act by structurally mimicking the C-terminal cyclic imide degron, which allows them to bind CRBN and redirect it to target novel "neosubstrates" for proteasomal degradation [1]. Recent efforts have employed prodrugs that exploit a mechanism analogous to endogenous intramolecular cyclization to generate cyclic imide-based drugs in situ within cells [4]. Cyclic imide-based warheads inspired by the degron have also been used as scaffolds for proteolysis-targeting chimeras (PROTACs) to enable the degradation of protein targets with high potency and selectivity [5, 6]. The ability to generate and measure these species is, therefore, of interest for both fundamental research and translational applications.

To support investigation of CRBN substrates, efficient methods are needed to generate peptides or proteins bearing the C-terminal cyclic imide modification. In this chapter, we describe protocols for the controlled generation of C-terminal cyclic imide-modified peptides and proteins using two complementary approaches: (i) sortase-mediated transpeptidation, which enables tagging an engineered protein of interest with a synthetic peptide bearing cN; and (ii) enzymatic conversion catalyzed by PCMT1, which promotes cN formation at C-terminal asparagine residues on peptide or protein substrates [3]. In addition, we describe analytical strategies for verifying C-terminal cyclic imide formation and assessing its functional consequences, including time-resolved Förster resonance energy transfer (TR-FRET) assays and mass spectrometry (MS). Additional methods for detecting cyclic imide-modified proteins in complex biological samples or for characterizing cyclic imide-based small-molecule ligands by TR-FRET are available in related work [7, 8]. Together, these techniques provide a toolkit for probing the chemistry and biology of the C-terminal cyclic imide.

2 Materials

2.1 Chemical Synthesis of C-Terminal Cyclic Imide-Bearing Peptides

1. 2-chlorotrityl chloride (2-CTC) resin.
2. Hexafluoroisopropanol.
3. Dichloromethane (DCM).
4. Acetonitrile (MeCN).
5. Dimethylformamide (DMF).
6. HCl•cN (in house).
7. Piperidine.

8. Dimethylisopropylethylamine (DIPEA).
9. Hexafluorophosphate azabenzotriazole tetramethyl uronium (HATU).
10. Triisopropylsilane (TIPS).
11. Ethanedithiol (EDT).
12. Trifluoroacetic acid (TFA).

2.2 Sortase-Mediated Conjugation of C-Terminal Cyclic Imides onto Target Proteins

1. pET29-eSrtA.
2. Co-NTA magnetic beads.
3. Zeba spin desalting columns, 7K MWCO, 5 mL.
4. Vivaspin spin concentrators.

2.3 PCMT1-Mediated Generation of C-Terminal Cyclic imides

1. *S*-adenosyl-methionine (SAM).
2. Tris(hydroxymethyl)aminomethane (Tris).

2.4 TR-FRET Characterization of C-Terminal Cyclic Imide Species

1. Tween-20.
2. Triton X-100.
3. Proxiplate-384 Plus.
4. Anti-His_6 antibody.
5. Zeba spin desalting columns, 7K MWCO, 0.5 mL.
6. C5 lenalidomide (CAS: 191732-70-4).
7. CoraFluor-1-pentafluorophenyl (Pfp) ester.
8. Thalidomide-FITC.
9. D300e digital dispenser (HP).
10. Plate reader compatible with TR-FRET measurements, such as PHERAstar FSX microplate reader (BMG Labtech).

2.5 MS Characterization of C-Terminal Cyclic Imide Species

1. S-Trap micro columns.
2. Pierce peptide desalting spin columns.
3. Proteases (dependent on protein of interest).
4. High-resolution LC–MS system with electrospray ionization, such as MicrOTOF II LC–MS (Bruker) for peptides, Impact II qTOF (Bruker) for intact proteins, and Orbitrap Eclipse Tribrid Mass Spectrometer (Thermo Fisher) for peptide digests.

3 Methods

3.1 Chemical Synthesis of C-Terminal Cyclic Imide-Bearing Peptides

1. Prepare the protected peptide by standard Fmoc-based solid-phase peptide synthesis (SPPS) on 2-chlorotrityl chloride (2-CTC) resin (loading 1.5–2.0 mmol/g, 0.25 mmol scale, Fig. 1a) [9, 10].
2. Cleavage: Add ~5 mL hexafluoroisopropanol (HFIP)/dichloromethane (DCM) (1:4 v/v) to the peptide on resin and shake for 30 min at 24 °C. Filter the mixture, wash the resin with 10 mL DCM, and reduce the filtrate to minimal volume under nitrogen. Dissolve the crude protected peptide in 10 mL acetonitrile (MeCN)/water (1:1 v/v), further dilute to ~25% MeCN with water, and lyophilize to dryness (Fig. 1b).
3. C-terminal cyclic imide coupling: Dissolve the crude protected peptide (0.10 mmol, 1.0 equiv) and HCl•cN (0.30 mmol, 3.0 equiv) in 2.0 mL dimethylformamide (DMF, 0.05 M). Add dimethylisopropylethylamine (DIPEA; 0.60 mmol, 6.0 equiv) and HATU (0.30 mmol, 3.0 equiv) sequentially while stirring at 24 °C. After 6 h, remove DMF using a rotary evaporator (Fig. 1c).
4. Deprotection: Dissolve the crude material in 2 mL of 50% trifluoroacetic acid (TFA)/DCM containing 3% triisopropylsilane (TIPS). For peptides containing Met, add 3% ethanedithiol to this solution. After stirring at 24 °C for 1.5 h, reduce the mixture to minimal volume under nitrogen, then add 1 mL DMF. Purify by semi-preparative HPLC to yield the desired C-terminal cyclic imide-bearing peptide (Fig. 1d).

3.2 Sortase-Mediated Conjugation of C-Terminal Cyclic Imides onto Target Proteins

1. Express the sortase enzyme using the pET29-eSrtA [11] plasmid and purify by Ni-NTA and size exclusion chromatography as previously described [1]. Dilute the purified sortase to 100 μ M, flash-freeze in small aliquots, and store at −80 °C until use.
2. Design, express, and purify the target protein terminating in the sequence ...GGSLPETGHHHHHH. Flash-freeze in small aliquots and store at −80 °C until use.
3. Prepare the following sortase reaction mixture (Fig. 2, *see* **Note 2**):

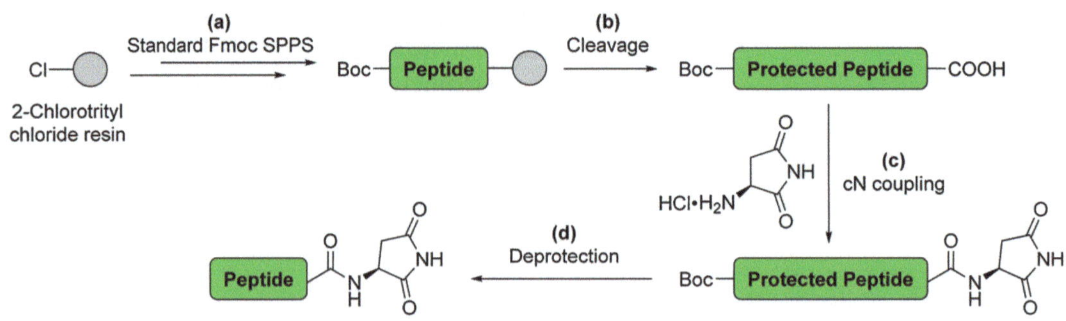

Fig. 1 Scheme of chemical synthesis to generate peptides bearing the C-terminal cyclic imide

   a. 10 μM target protein.
   b. 0.4 mM GGG peptide bearing the C-terminal cN.
   c. Buffer: 100 mM Tris–Cl, 150 mM NaCl, 5 mM $CaCl_2$, pH 7.5.
   d. 2 μM eSrtA (add last).
4. Incubate the reaction mixture at 24 °C for 1 h.
5. Add 25 μL of prewashed Co-NTA magnetic beads to capture and remove unreacted His_6-tagged protein and sortase. Incubate at 24 °C for 15 min with gentle inversion.
6. Collect the beads magnetically and recover the supernatant containing the conjugated protein.
7. To remove residual peptides, exchange the buffer to PBS using Zeba columns according to the manufacturer's protocol.
8. Concentrate the conjugated protein using Vivaspin centrifugal filters of the appropriate MWCO.
9. Determine protein concentration by measuring absorbance at 280 nm.
10. Confirm conjugation by SDS-PAGE and/or intact mass spectrometry.
11. Flash-freeze the conjugated protein in small aliquots and store at −80 °C.

3.3 PCMT1-Mediated Generation of C-Terminal Cyclic Imides

1. Prepare a reaction mixture containing PCMT1 (10–20 μM, 1.0–2.0 equiv) and *S*-adenosyl-methionine (SAM; 200 μM, 20 equiv) in 50 mM Tris–Cl buffer, pH 7.4 (*see* **Note 3**). Warm the mixture to 37 °C for 5 min. Add the peptide or protein substrate bearing a C-terminal asparagine (10 μM) and incubate at 37 °C for 12–24 h (Fig. 3, *see* **Note 4**).

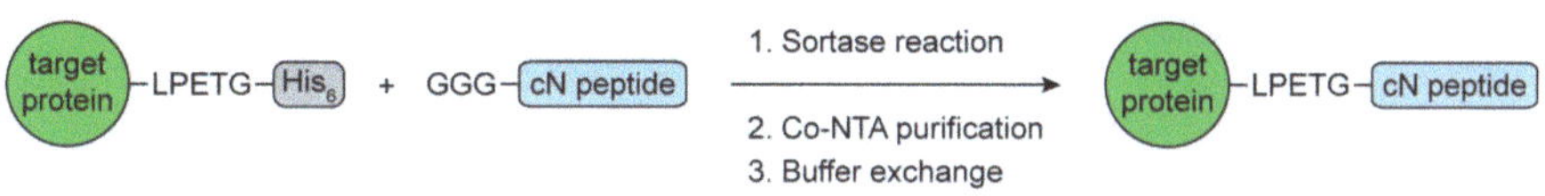

Fig. 2 Scheme of sortase-mediated conjugation of a C-terminal cyclic imide onto a target protein

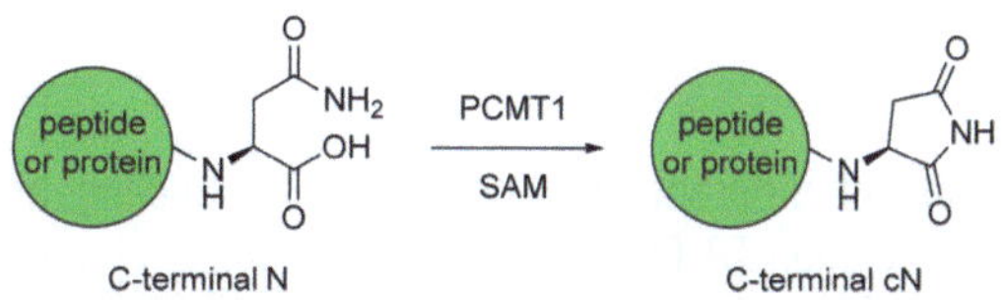

Fig. 3 Scheme of PCMT1-mediated C-terminal cN formation on peptides or proteins

2. If necessary, purify peptides by reverse-phase HPLC or proteins by FPLC to remove PCMT1 and excess SAM. The isolated product typically contains a mixture of C-terminal N and cN species.

3.4 TR-FRET Characterization of C-Terminal Cyclic Imide Species

3.4.1 Preparation of TR-FRET Reagents

1. Purify His_6-CRBN–DDB1 as previously described [12].
2. Label either the anti-His_6 antibody or the CRBN–DDB1 complex directly with the terbium-based CoraFluor-1 (henceforth, referred to as Tb; Fig. 4A). Exchange the protein into labeling buffer (100 mM sodium carbonate, 0.05% Tween-20 [v/v], pH 8.5) using Zeba columns to remove amine-containing components (*see* **Note 5**).
3. Perform Tb labeling as described previously [7, 13]. Add CoraFluor-1-pentafluorophenyl (Pfp) ester at a threefold molar excess relative to the protein.

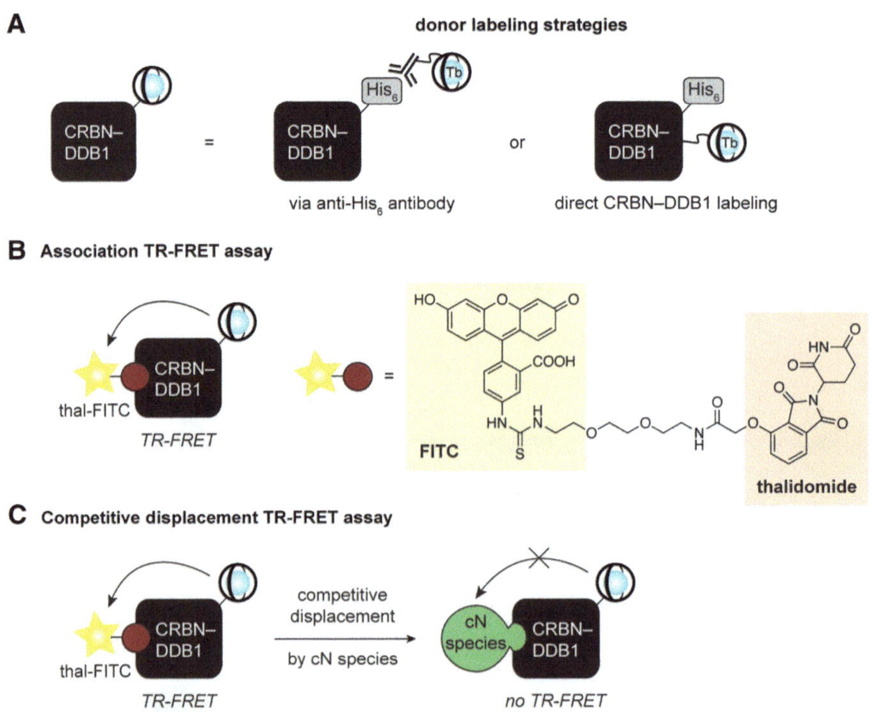

Fig. 4 (**A**) Scheme of different approaches for Tb labeling of His_6-CRBN–DDB1 to generate the donor. (**B**) Scheme of association TR-FRET assay to measure the K_D of the tracer, and the structure of the thalidomide-FITC tracer. (**C**) Scheme of competitive displacement TR-FRET assay to measure the affinity of C-terminal cyclic imide species to CRBN–DDB1

4. Gently mix the reaction by vortexing and incubate at 24 °C for 1 h.
5. Quench the reaction by exchanging the mixture into storage buffer (50 mM sodium phosphate, 150 mM NaCl, 0.05% Tween-20, pH 7.4) using Zeba columns.
6. Determine the final concentration of the Tb-labeled protein by correcting the absorbance at 280 nm according to,

$$A_{280,corr} = A_{280} - 0.157 \times A_{340}$$

where A_{280} and A_{340} are the absorbances at 280 nm and 340 nm, respectively [8]. Convert $A_{280,corr}$ to protein concentration using the extinction coefficients ε_{280} = 210,000 $M^{-1}cm^{-1}$ for anti-His_6 antibody and ε_{280} = 167,000 $M^{-1}cm^{-1}$ for CRBN–DDB1 (*see* **Note 6**).
7. Flash-freeze the labeled protein in small aliquots and store at –80 °C.

3.4.2 Association TR-FRET Assay to Determine Tracer K_D

1. Perform an association TR-FRET assay to determine the dissociation constant (K_D) of the thalidomide-FITC tracer (Fig. 4B). Prepare sufficient reaction volume for triplicate measurements of each condition. Mix the following components at the indicated final concentrations:
   a. 5 nM Tb-labeled anti-His_6 antibody and 10 nM His_6-CRBN–DDB1, or 10 nM Tb-labeled His_6-CRBN–DDB1 (*see* **Note 7**).
   b. Buffer: 25 mM HEPES, 150 mM NaCl, 0.5 mg/mL BSA, 0.005% (v/v) Tween-20, pH 7.5.
2. Gently nutate the mixture at 24 °C for at least 5 min.
3. Dispense 15 μL of the mixture per well into a low-volume 384-well plate (e.g., Proxiplate-384 Plus).
4. Using an HP D300e digital dispenser, add thalidomide-FITC (from a DMSO stock) as a 1:2 serial dilution series (8 points, highest concentration C_{max} = 1.33 μM).
5. When using the Tb-labeled anti-His_6 antibody, include control wells containing the same concentration of Tb-labeled antibody and the thalidomide-FITC dilution series, but lacking His_6-CRBN–DDB1, to account for tracer-dependent background TR-FRET signal. When using Tb-labeled His_6-CRBN–DDB1, include no-tracer wells for signal normalization.
6. Centrifuge the plate at 500 × *g* for 1 min, then allow it to equilibrate at 24 °C for 10 min before measurement.

7. Record TR-FRET signals on a plate reader with the following parameters: excitation at 350 nm; emission at 616 nm (Tb) and 520 nm (FITC); 110 flashes per read; 0.05 ms excitation time; 0.05 ms delay; and 0.2 ms integration time.
8. Calculate the TR-FRET ratio for each well as the 520/616 nm emission intensity ratio. Normalize signals by subtracting the background from the corresponding wells and determine the apparent tracer dissociation constant ($K_{D,app}$) by fitting to a one-site binding model.
9. If using Tb-labeled His_6-CRBN–DDB1, the true tracer dissociation constant ($K_{D,trac}$) is equal to $K_{D,app}$. If using the Tb-labeled anti-His_6 antibody, a further correction is required to adjust for the bivalency of the antibody:

$$K_{D,trac} = K_{D,app} \times \left(1 + \sqrt{2}\right)$$

10. For all subsequent displacement assays, include the tracer in each well at a fixed concentration, typically 0.3–0.5 × $K_{D,trac}$.

3.4.3 Displacement TR-FRET Assay

1. Prepare C-terminal cyclic imide species at a final concentration of 10 μM. Peptides, purified proteins, or crude PCMT1 incubation mixtures (initial substrate concentration 10 μM) can be used (*see* **Note 8**).
2. Add Triton X-100 to a final concentration of 0.1 % (v/v) to improve dispensing accuracy.
3. Perform a displacement TR-FRET assay (Fig. 4C) to determine the affinity of C-terminal cyclic imide species to CRBN–DDB1. Prepare sufficient reaction volume for triplicate measurements of each condition. Mix the following components at the indicated final concentrations:
   a. 5 nM Tb-labeled anti-His_6 antibody and 10 nM His_6-CRBN–DDB1, or 10 nM Tb-labeled His_6-CRBN–DDB1.
   b. Buffer: 25 mM HEPES, 150 mM NaCl, 0.5 mg/mL BSA, 0.005% (v/v) Tween-20, pH 7.5.
   c. Thalidomide-FITC at a fixed concentration selected based on its measured $K_{D,trac}$ (typically 0.3–0.5 × $K_{D,trac}$).
4. Gently nutate the mixture at 24 °C for at least 5 min.
5. Dispense 15 μL of the mixture per well into a low-volume 384-well plate (e.g., Proxiplate-384 Plus).
6. Using an HP D300e digital dispenser, add the C-terminal cyclic imide species as a 1:3 serial dilution series (8 points, highest concentration C_{max} = 1 μM). Adjust dilution range as necessary to capture full displacement curves.

7. Include control wells for signal normalization:
   a. Wells containing all components except the C-terminal cyclic imide competitor species (maximal FRET, "top").
   b. Wells containing 100 μM C5 lenalidomide instead of the C-terminal cyclic imide species (maximal competition and minimal FRET, "bottom").
8. Centrifuge the plate at 500 × *g* for 1 min, then allow it to equilibrate at 24 °C for 10 min before measurement.
9. Record TR-FRET measurements using the same instrument parameters described above.
10. Calculate the TR-FRET ratio for each well as the 520/616 nm emission intensity ratio. Subtract the assay bottom from all values, normalize to the assay top, and fit normalized data (0–1 range) to a four-parameter logistic model [log(inhibitor) vs. response – variable slope] with Top = 1 and Bottom = 0 to obtain IC_{50} values.
11. Calculate the K_i of the C-terminal cyclic imide species from its IC_{50} value using the Cheng-Prusoff relationship:

$$K_i = \frac{IC_{50}}{1 + \frac{[trac]}{K_{D,trac}}},$$

where [trac] is the concentration of tracer.

3.5 MS Characterization of C-Terminal Cyclic Imide Species

3.5.1 MS Characterization of Peptides

1. Mix 2.5 μL of the C-terminal cyclic imide-bearing peptide sample (purified or crude PCMT1 reaction mixture, 10 μM) with 20 μL MeCN. Add formic acid (FA) to a final concentration of 0.1%.
2. Analyze the mixture by LC–MS using a high-resolution electrospray ionization (ESI) mass spectrometer.
3. Extract ion chromatograms corresponding to the acyclic, methyl ester, and C-terminal cyclic imide forms of the peptide to assess the efficiency of cyclic imide formation (*see* **Note 9**).

3.5.2 Intact MS Characterization Proteins (see Note 10)

1. Add FA to a final concentration of 0.1% to the C-terminal cyclic imide-bearing protein sample (purified or crude PCMT1 reaction mixture, 1–20 μM).
2. Analyze the mixture by LC–MS using high-resolution ESI conditions suitable for intact proteins.

3. Analyze raw spectra by maximum-entropy deconvolution to compare the relative levels of acyclic, methyl ester, and C-terminal cyclic imide species.

3.5.3 Bottom-Up MS Characterization of Proteins

1. To a 100 μL sample of C-terminal cyclic imide-bearing protein (purified or crude PCMT1 reaction mixture), add 100 μL of 10% (w/v) sodium dodecyl sulfate (SDS) in 100 mM Tris–Cl, pH 7.0.
2. Add 20 μL of 27.5% phosphoric acid and vortex briefly to acidify the sample.
3. Add 1.32 mL of S-Trap binding buffer (100 mM triethylammonium bicarbonate in 90% methanol, pH 7.1) and mix well. Do not centrifuge at this stage.
4. Load the mixture onto an S-Trap microcolumn under vacuum. Wash three times with 1 mL of binding buffer, then centrifuge at 4000 × *g* for 1 min to remove residual solvent.
5. Digest proteins on-column with an appropriate protease following the manufacturer's guidelines (*see* **Note 11**). Keep the incubation as brief and cool as possible to minimize cyclic imide hydrolysis.
6. Sequentially elute peptides by adding 40 μL of the following, centrifuging at 4000 × *g* for 1 min after each addition:
   a. 100 mM ammonium bicarbonate, pH 7.0.
   b. 0.2% FA.
   c. 0.2% FA in 50% MeCN.

   Pool eluates, dry in a vacuum concentrator, and resuspend in 300 μL of 0.1% TFA.
7. Desalt samples using Pierce peptide desalting spin columns according to the manufacturer's protocol. Dry in a vacuum concentrator and resuspend in 20 μL of 0.1% FA.
8. Inject 1 μL of the peptide sample onto a C18 analytical column coupled to a high-resolution LC–MS system equipped with an ESI source.
9. Separate peptides using a linear gradient from 1% to 40% MeCN with 0.1% FA.
10. Acquire MS data in positive ion mode in full-scan or selected ion monitoring mode for expected C-terminal cyclic imide-bearing peptides. Monitor both the unmodified and cyclic imide forms of peptides.
11. Extract ion chromatograms for each target peptide, integrate peak areas, and compare relative levels of unmodified and cyclic imide-bearing peptides (*see* **Note 12**).

4 Notes

1. Although both cN and cQ can form, our work to date has identified the C-terminal cN as the more physiologically relevant modification [2, 3]. Therefore, the protocols presented here focus on cN. However, all methods in this work, except for PCMT1-mediated formation, are compatible with measurement of cQ as well.
2. The C-terminal cyclic imide is hydrolytically unstable in aqueous solutions ($t_{1/2} \approx 24\,h$) [1], especially under basic conditions (pH ≥8.0). Whenever a C-terminal cyclic imide species-containing is present, work quickly and on ice when possible, and avoid exposure to basic buffers.
3. The efficiency of PCMT1-mediated cyclic imide formation varies depending on the substrate. The amount of PCMT1 stated in this protocol is recommended as a starting point. If higher conversion is required, use more equivalents of PCMT1.
4. The optimal PCMT1 incubation time may vary depending on the substrate, as C-terminal cyclic imide formation and hydrolysis occur concurrently.
5. Ensure that all Tb labeling steps are performed in amine-free buffer; Tris interferes with the Pfp ester coupling reaction [8].
6. The degree of Tb labeling can be estimated spectrophotometrically by calculating and comparing the concentrations of Tb, determined from A_{340} using $\varepsilon_{340} = 22{,}000\,M^{-1}cm^{-1}$, and the labeled protein (determined using $A_{280,corr}$ as described above). Calculate the labeling ratio as [Tb]/[protein].
7. The same method of Tb labeling should be used throughout all TR-FRET experiments to ensure consistency.
8. If using Tb-labeled anti-His_6 antibody as the TR-FRET donor, make sure that CRBN–DDB1 is the only protein in the mixture that has a His_6 tag.
9. The expected mass shift relative to the unmodified species is +14.0157 Da for the methyl ester and −18.0106 Da for the C-terminal cyclic imide.
10. When characterizing C-terminal cyclic imide-bearing proteins by MS, choose intact MS for rapid verification of cyclic imide formation and bottom-up MS for definitive identification, as identical mass shifts can arise from other PTMs on the protein (e.g., dehydration).
11. The protease chosen for bottom-up MS depends on the specific protein being analyzed. Choose a protease that produces

a C-terminal peptide of optimal length for MS analysis (typically, 8–15 amino acids). For example, avoid using trypsin if there is a K or R residue too proximal to the C-terminus.

12. When quantifying C-terminal cyclic imide formation by bottom-up MS, expect underestimation due to partial hydrolysis during sample preparation.

Acknowledgments

We thank M. Chen and S. Trager from the Harvard University Mass Spectrometry and Proteomics Resource Laboratory for support with protein MS experiments and D. Cui and B. Tresco from the Harvard University Laukien-Purcell Instrumentation Center for support with peptide MS experiments. Support from the National Institutes of Health (R01GM141406), the Blavatnik Biomedical Accelerator at Harvard University, Mark Foundation for Cancer Research (C.M.W.), Merkin Family Foundation, and the Starr Foundation is gratefully acknowledged.

References

1. Ichikawa S, Flaxman HA, Xu W, Vallavoju N, Lloyd HC, Wang B, Shen D, Pratt MR, Woo CM (2022) The E3 ligase adapter cereblon targets the C-terminal cyclic imide degron. Nature 610(7933):775–782. https://doi.org/10.1038/s41586-022-05333-5
2. Xu W, Zhao Z, Su M, Jain AD, Lloyd HC, Feng EY, Cox N, Woo CM (2025) Genesis and regulation of C-terminal cyclic imides from protein damage. Proc Natl Acad Sci U S A 122(1):e2415976121. https://doi.org/10.1073/pnas.2415976121
3. Zhao Z, Xu W, Feng EY, Cao S, Hermoso-Lopez A, Pena-Vega P, Lloyd HC, Porter AKD, Guzman M, Zheng N, Woo CM (2026) PCMT1 generates the C-terminal cyclic imide degron on CRBN substrates. Nat Chem Biol 22(5):802–812. https://doi.org/10.1038/s41589-025-02106-9
4. Chase DH, Stein A, Grinshpun DE, Krone MW, Crews CM (2025) Design and application of cereblon-recruiting prodegraders. J Am Chem Soc 147(28):24527–24537. https://doi.org/10.1021/jacs.5c05036
5. Ichikawa S, Payne NC, Xu W, Chang CF, Vallavoju N, Frome S, Flaxman HA, Mazitschek R, Woo CM (2024) The cyclimids: degron-inspired cereblon binders for targeted protein degradation. Cell Chem Biol 31(6):1162–1175. https://doi.org/10.1016/j.chembiol.2024.01.003
6. Zhu Q, Fischer G, Cheng SS, Payne NC, Peter D, Mody AC, Arce-Solano S, Shen D, Lin Z, Mazitschek R, Kessler D, Woo CM (2025) Enzyme-activated sugar-coated bifunctional degraders. J Am Chem Soc 147(38):34672–34680. https://doi.org/10.1021/jacs.5c09843
7. Lloyd HC, Li Y, Payne NC, Zhao Z, Xu W, Kroupova A, Zollman D, Long T, Kabir F, Chen M, Freeman R, Feng EY, Xi SY, Hsu Y-C, Ciulli A, Mazitschek R, Woo CM (2025) A method for the detection and enrichment of endogenous cereblon substrates. Cell Chem Biol 32(8):1028–1041. https://doi.org/10.1016/j.chembiol.2025.07.002
8. Payne NC, Ichikawa S, Woo CM, Mazitschek R (2024) Protocol for the comprehensive biochemical and cellular profiling of small-molecule degraders using CoraFluor TR-FRET technology. STAR Protoc 5(2):103129. https://doi.org/10.1016/j.xpro.2024.103129
9. Fields GB, Noble RL (1990) Solid phase peptide synthesis utilizing 9-fluorenylmethoxycarbonyl amino acids. Int J Pept Protein Res 35(3):161–214. https://doi.org/10.1111/j.1399-3011.1990.tb00939.x
10. Chatzi KBO, Gatos D, Stavropoulos G (2009) 2-Chlorotrityl chloride resin. Int J Pept Protein Res 37(6):513–520. https://doi.org/10.1111/j.1399-3011.1991.tb00769.x

11. Chen I, Dorr BM, Liu DR (2011) A general strategy for the evolution of bond-forming enzymes using yeast display. Proc Natl Acad Sci U S A 108(28):11399–11404. https://doi.org/10.1073/pnas.1101046108
12. Lin W, Li Y, Min J, Liu J, Yang L, Lee RE, Chen T (2020) Development of BODIPY FL thalidomide as a high-affinity fluorescent probe for cereblon in a time-resolved fluorescence resonance energy transfer assay. Bioconjug Chem 31(11):2564–2575. https://doi.org/10.1021/acs.bioconjchem.0c00507
13. Payne NC, Kalyakina AS, Singh K, Tye MA, Mazitschek R (2021) Bright and stable luminescent probes for target engagement profiling in live cells. Nat Chem Biol 17(11):1168–1177. https://doi.org/10.1038/s41589-021-00877-5

Chapter 6

Design and Semisynthesis of Ubiquitin Extension Probes

Chuntong Li, Luyu Shi, Yingyue Zhang and Jinghong Li

Abstract

Activity-based ubiquitin probes serve as crucial tools for capturing and elucidating the molecular mechanisms of E3 ligase-catalyzed substrate ubiquitination. We recently reported a chemoenzymatic strategy for generating ubiquitin chain extension probes, which offers advantages such as mild reaction conditions, simple operation, and high specificity, thereby providing new opportunities for E3 enzyme research. This article details the experimental protocol for obtaining a K48-linked di-ubiquitin extension probe based on a substrate-Type II degron using the chemoenzymatic approach.

Keywords Ubiquitin extension probes, Chemoenzymatic strategy, Ubiquitin E3 ligase, Ubiquitin chain modification, Type II degron

1 Introduction

As a pivotal post-translational modification in eukaryotes, ubiquitination regulates nearly all biological processes, including the cell cycle and DNA damage repair [1–3]. The ubiquitination process is coordinately executed by the sequential actions of E1, E2, and E3 enzymes [4]. Among these, ubiquitin E3 ligases directly recognize and ubiquitinate substrate proteins, playing a central role in determining substrate localization, degradation, and protein–protein interactions [5]. With over 600 distinct E3 ligases encoded in the human genome, dysregulation of their activity is closely associated with various diseases [6–8]. For instance, overexpression of the E3 ligase MDM2 leads to suppressed levels of the tumor suppressor p53 [9], thereby promoting uncontrolled proliferation of tumor cells. Additionally, mutations in the Parkin gene, which encodes another E3 ligase, represent one of the most common causes of familial Parkinson's disease [10–12]. Within

The authors "Chuntong Li and Luyu Shi" contributed equally.

Maria W. Górna et al. (eds.), *Degrons*, Methods in Molecular Biology, vol. 3069,
https://doi.org/10.1007/978-1-0716-5508-5_6,

this context, the development of molecules that target or even hijack E3 ligase activity has attracted increasing attention in biomedical research [13–15].

The ubiquitination of substrates by E3 enzymes is characterized by its transient and dynamic nature, presenting a longstanding challenge for mechanistic dissection. In recent years, activity-based ubiquitin probes that covalently capture transient intermediates during the enzymatic process have emerged as powerful tools for studying E3 ligases [10, 16–24]. E3 ligases are primarily classified into RING, HECT, and RBR types, with RING-type E3s being the most prevalent [5]. Since RING E3 ligases lack a catalytic cysteine residue, strategies utilizing chemically synthesized E2–Ub–substrate conjugates, termed ubiquitin extension probes, have been developed [24–32]. These probes are incubated with E3 ligases to mimic the transient ubiquitination intermediate, enabling mechanistic studies. This approach has been successfully applied to elucidate the molecular mechanisms of various RING E3 ligases, for instance, the UBR1-mediated type I N-degron ubiquitination initiation and K48-linked di-ubiquitin chain extension [24], and the nucleosome-associated E3 Bre1 catalyzed H2A K119 mono-ubiquitination [33]. Collectively, these strategies have significantly advanced our understanding of substrate recognition and ubiquitin chain formation by RING-type E3 ligases.

Recently, we reported a chemoenzymatic strategy for generating E2–Ub chain extension conjugates [22], offering a new approach to investigate the process of E3 ligase-catalyzed substrate polyubiquitination. Here, we designed and synthesized an intermediate mimic that recapitulates the E3 ligase Ubr1 in complex with the E2 enzyme UBC2 to catalyze K48-linked di-ubiquitin formation on a type II degron substrate (Fig. 1). This strategy first involves E1-mediated native chemical ligation [34–36] between Ub-thioester and CAET (2-((2-Chloroethyl)amino)ethane-1-thiol) modified Ub at position 48 to generate CAET-linked K48-diUb. Subsequently, the E2 enzyme UBE2E1 [37] directly conjugates the CAET-linked K48-diUb to the ubiquitination site of the type II degron substrate. Finally, activation of the catalytic cysteine residue on the E2 enzyme enables disulfide bond formation with the thiol group of the CAET molecule, allowing the di-Ub extension probe to be obtained under physiological conditions.

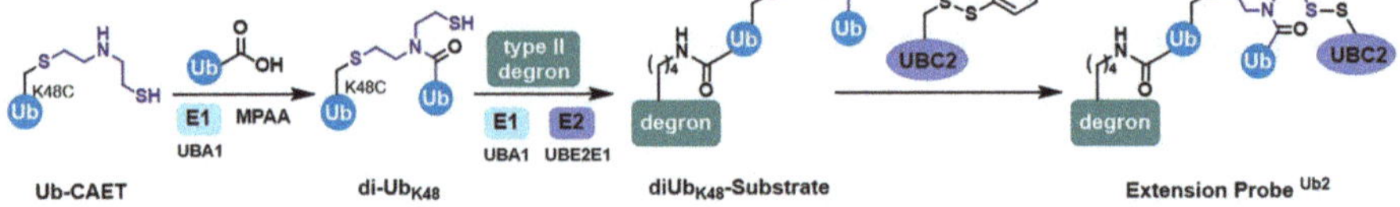

Fig. 1 Synthetic route for extension probe^{Ub2}

2 Materials

All solutions are prepared using ultra-pure water (dd H_2O, 18 MΩ·cm). Adjust pH using 2 M NaOH and 2 M HCl solutions.

2.1 Protein Expression and Purification

I. Bacterial Strains and Culture Media

1. LB Broth Powder.
2. 9 cm Bacterial Petri Dishes (LB Agar).
3. BL21 (DE3) Chemically Competent Cells.

II. Antibiotics

4. Ampicillin Stock Solution (100 mg/mL).
5. Kanamycin Stock Solution (50 mg/mL).

III. Protein Expression Inducers and Inhibitors

6. Isopropyl β-D-thiogalactoside (IPTG) Stock Solution (1 M).
7. Phenylmethylsulfonyl Fluoride (PMSF) Stock Solution (100 mM in isopropanol).

IV. Buffers and Solutions

8. 4-(2-Hydroxyethyl)-1-piperazineethanesulfonic acid (HEPES) Solution (1 M, pH 8.0).
9. Sodium Acetate (NaOAc) Solution (1 M, pH 4.5).
10. Sodium Phosphate Monobasic (NaH_2PO_4) Solution (1 M, pH 7.5).
11. Sodium Chloride (NaCl) Solution (5 M).
12. Imidazole Solution (5 M, pH 7.5).
13. Mobile phase A for HPLC: Acetonitrile (ACN) (0.1% TFA): Add 5 mL TFA to 5 L ACN.
14. Mobile phase B for HPLC: ddH_2O (0.1% TFA): Add 5 mL TFA to 5 L ddH_2O.

V. Chromatography Columns and Resins

15. HisTrap HP Column (5 mL).
16. Capto HiRes S Column (10/100).
17. Superdex™ 75 Increase 10/300 GL Column.
18. Superdex™ 200 Increase 10/300 GL Column.
19. XB-C4 HPLC analytical column XB-C4 (4.6×250 mm, 5 μm).
20. XB-C4 HPLC semipreparative column (10×250 mm, 5 μm).

VI. Laboratory Consumables

21. Microcentrifuge Tubes (1.5 mL).
22. Conical Centrifuge Tubes (50 mL).
23. Plate spreader.
24. Dialysis Tubing (3.5 kDa MWCO).
25. Syringe Filters (0.22 μm).

VII. Instruments and Equipment

26. pH Meter.
27. Thermostat Water Bath.
28. Ultrasonic Cell Disruptor.
29. Ultracentrifuge.
30. Thermomixer/Mixing Incubator.
31. Low-volume spectrophotometer such as NanoDrop™.
32. Gel Electrophoresis System.
33. Protein Chromatography System such as Cytiva ÄKTA pure 25.
34. High-performance liquid chromatograph such as Shimadzu LC-20.
35. High-performance liquid chromatograph mass spectrometer such as Shimadzu LCMS-2050.
36. Lyophilizer such as Eyela FDU-2110.

2.2 Synthesis of K48-Linked Diubiquitin via Chemical Ligation

I. Chemicals and Reagents

1. 4-Mercaptophenylacetic acid (MPAA).
2. 2-((2-Chloroethyl)amino)ethane-1-thiol acetamidomethyl (CAET-Acm).
3. Palladium chloride ($PdCl_2$).
4. Guanidine hydrochloride (Gn·HCl), 6 M.
5. Tris(2-carboxyethyl)phosphine (TCEP).
6. Dithiothreitol (DTT).
7. Adenosine 5′-triphosphate disodium salt trihydrate (Na_2 ATP·3 H_2O) Solution (0.5 M, pH 6.2).
8. Magnesium chloride hexahydrate ($MgCl_2 \cdot 6\, H_2O$) Solution (1 M, pH 6.2).

II. Buffers and Solutions

9. Conjugation Buffer: 6 M Gn·HCl, 0.1 M HEPES, 10 mM TCEP, pH 8.5.

10. HPLC Dilution Buffer: 6 M Gn·HCl, 0.2 M Na_2HPO_4, pH 8.0.
11. Deprotection Buffer: 6 M Gn·HCl, 0.2 M Na_2HPO_4, pH 7.5.
12. Refolding Buffer: 4 M Gn·HCl, 0.2 M Na_2HPO_4, 1 mM TCEP, pH 7.5.
13. Dialysis Buffer 1: 1 M Gn·HCl, 0.2 M Na_2HPO_4, 1 mM TCEP, pH 7.5.
14. Dialysis Buffer 2: 0.5 M Gn·HCl, 0.2 M Na_2HPO_4, 1 mM TCEP, pH 7.5.
15. SEC Buffer: 50 mM HEPES, 150 mM NaCl, pH 7.5.
16. Reaction Buffer (Diubiquitin Formation): 20 mM HEPES, 150 mM NaCl, 5 mM TCEP, 10 mM ATP, 10 mM $MgCl_2$, 25 mM MPAA, pH 7.5.

III. Proteins/Enzymes:

17. Ubiquitin K48C mutant (Ub K48C), lyophilized.
18. Ubiquitin (Ub), wild-type (5 mg/mL).
19. Human E1 enzyme (hUBA1, 1 μM).

2.3 Enzymatic Synthesis of Ubiquitinated Substrate

I. Proteins/Peptides:

1. SUE1-tagged type II degron peptide.
2. K48-diUb-CAET (100 μM).
3. Human E2 enzyme UBE2E1 (100 μM).
4. Human E1 enzyme hUBA1 (2 μM).

II. Buffers and Solutions

5. Reaction Buffer: 20 mM HEPES, 150 mM NaCl, 5 mM TCEP, 10 mM ATP, 5 mM $MgCl_2$, pH 7.5.

2.4 Construction of the Ub Extension Probe

I. Chemicals and Reagents

1. 5,5'-Dithiobis-(2-nitrobenzoic acid) (DTNB).

II. Buffers and Solutions

2. Reaction Buffer: 20 mM HEPES, 150 mM NaCl, pH 7.5.

III. Proteins/Enzymes:

3. UBC2 (3 mg).
4. K48-diUb-CAET-degron Conjugate (K48-diUb-CAET-Sub) (0.5 mg).

3 Methods

3.1 Protein Expression and Purification

3.1.1 Expression and Purification of Ubiquitin (Ub)

1) The codon-optimized gene encoding full-length human Ub was synthesized by Genscript (Nanjing, China) and cloned into a pET22b vector.
   The amino acid sequence is:
   MQIFVKTLTGKTITLEVEPSDTIENVKAKIQDKEGIPPD-QQRLIFAGKQLEDGRTLSDYNIQKESTLHLVLRLRGG.
2) Transform the recombinant plasmid into *E. coli* BL21 (DE3) competent cells. Inoculate a single colony into 20 mL of fresh LB medium supplemented with 50 μg/mL ampicillin and culture overnight at 37°C.
3) Transfer the 20 mL preculture into 1 L of fresh LB medium containing 50 μg/mL ampicillin. Grow the culture at 37°C until the OD_{600} reaches 0.8. Induce protein expression by adding isopropyl β-D-thiogalactoside (IPTG) to a final concentration of 1 mM, and continue incubation for 6 h at 37°C.
4) Harvest the cells by centrifugation at 4000 × *g* for 30 min at 4°C. Resuspend the cell pellet in distilled H_2O and lyse by ultrasonication on ice for 60 min (using 100 W power with cycles of 3 s sonication followed by 6 s rest).
5) Add perchloric acid (70% v/v) to the lysate to a final concentration of 1.0% (v/v). Stir the mixture for 2 min and centrifuge at 12,000 × *g* for 30 min at 4°C.
6) Collect the supernatant and dialyze it overnight at 4°C against dialysis buffer (0.1% trifluoroacetic acid in H_2O) using a 3.5 kDa molecular weight cut-off (MWCO) dialysis membrane (*see* **Note 1**).
7) Filter the dialyzed sample through a 0.22 μm filter and purify by reverse-phase high-performance liquid chromatography (RP-HPLC) on an analytical XB-C4 column. Apply a linear gradient from 20% to 60% acetonitrile (containing 0.1% TFA) over 30 min.
8) Identify fractions containing Ub by mass spectrometry to confirm the correct molecular weight. Pool the target fractions and lyophilize for 48 h. Store the resulting powder at −80°C.

3.1.2 Expression and Purification of Ub K48C Mutant

1) The codon-optimized gene encoding the Ub K48C mutant was synthesized by Genscript and cloned into a pET22b vector.
   The amino acid sequence is:

MQIFVKTLTGKTITLEVEPSDTIENVKAKIQDKEGIP-PDQQRLIFAGCQLEDGRTLSDYNIQKESTLHLVLRL-RGG

2) Perform the expression and purification of Ub K48C identically to the procedures described for wild-type Ub in Sect. 3.1.1, Steps 2–8.

3.1.3 Expression and Purification of UBA1

1) The codon-optimized gene encoding full-length human UBA1 with an N-terminal His_6-tag (sequence: HHHHHHGSGS-) was synthesized by Genscript and cloned into a pET22b vector.
2) Transform the plasmid into *E. coli* BL21 (DE3) cells. Inoculate 20 mL of LB medium containing 50 μg/mL ampicillin with a single colony and culture overnight at 16°C.
3) Dilute the preculture into 1 L of fresh LB medium with 50 μ g/mL ampicillin. When the OD_{600} reaches 0.8, induce protein expression by adding 0.4 mM IPTG and incubate the culture overnight at 16°C.
4) Harvest the cells by centrifugation at 4000 × *g* for 30 min at 4°C. Resuspend the pellet in lysis buffer (50 mM HEPES, 150 mM NaCl, pH 7.5) and lyse by ultrasonication.
5) Clarify the lysate by centrifugation at 12,000 × *g* for 30 min at 4°C. Load the supernatant onto a Ni-NTA affinity column.
6) Wash the column extensively with wash buffer (50 mM HEPES, 150 mM NaCl, 20 mM imidazole, pH 7.5). Elute the bound protein with elution buffer (50 mM HEPES, 150 mM NaCl, 300 mM imidazole, pH 7.0).
7) Further purify the eluted protein by size-exclusion chromatography (SEC) on a Superdex 200 Increase 10/300 GL column, pre-equilibrated with SEC buffer (50 mM HEPES, 150 mM NaCl, pH 7.5).
8) Pool the pure fractions of UBA1, flash-freeze in liquid nitrogen, and store at −80°C.

3.1.4 Expression and Purification of UBE2E1 Variant

1) The codon-optimized gene for human UBE2E1 variants with an N-terminal His_6-tag was synthesized by Genscript and cloned into a pET22b vector.

   The amino acid sequence is:

   HHHHHHGSGSLEVLFQGPNSKLLSTSAKRIQKELA-DITLDPPPNCSAGPKGDNIYEWRSTILGPPGSVYEG-GVFFLDITFTPEYPFKPPKVTFRTRIYHCNINSQGVI-CLDILRDNWSPALTISKVLLSICSLLTDCNPADPLVG-SIATQYMTNRAEHDRMARQWTKRYAT.
2) Express and initially purify the protein via Ni-NTA affinity chromatography as described for UBA1 in Sect. 3.1.3, Steps 2–7.

3) Subject the eluate from the Ni-NTA column to SEC on a Superdex 75 Increase 10/300 GL column, pre-equilibrated with SEC buffer (50 mM HEPES, 150 mM NaCl, pH 7.5).
4) Flash-freeze the purified UBE2E1 fractions in liquid nitrogen and store at −80°C.

3.1.5 Expression and Purification of UBC2

1) The codon-optimized gene for full-length yeast UBC2 with an N-terminal His_6-tag was synthesized by Genscript and cloned into a pET22b vector.
   The amino acid sequence is:
   HHHHHHGSGSLEVLFQGPMSTPARRRLMRDFKRM-KEDAPPGVSASPLPDNVMVWNAMIIGPADTPYEDGT-FRLLLEFDEEYPNKPPHVKFLSEMFHPNVYANGEI-CLDILQNRWTPTYDVASILTSIQSLFNDPNPASPAN-VEAATLFKDHKSQYVKRVKETVEKSWEDDMDDMD-DDDDDDDDDDDDEAD.
2) Express and purify UBC2 using the same protocol detailed for UBE2E1 in Sect. 3.1.4, Steps 2–4.

3.2 Synthesis of K48-Linked Diubiquitin via Chemical Ligation

3.2.1 Conjugation of Ub K48C with CAET-Acm (Steps 1–4)

1) Reaction Setup: Dissolve 24 mg of lyophilized Ub K48C in 880 μL conjugation buffer (6 M Gn·HCl, 0.1 M HEPES, 10 mM TCEP, pH 8.5) to achieve a 3 mM solution. Add 120 μL of 1 M CAET-Acm stock solution (final concentration: 120 mM). Adjust pH to 8.5 using 1 M NaOH.
2) Incubation: React at 37°C with shaking (120 rpm) for 10–14 h.
3) pH Monitoring: Measure pH every 2 h and readjust to 8.5 with NaOH as needed (*see* **Note 2**).
4) Reaction Monitoring: Analyze reaction progress by HPLC. Dilute 5 μL aliquots in 900 μL HPLC dilution buffer (6 M Gn·HCl, 0.2 M Na_2HPO_4, pH 8.0), inject via sample loop, and separate using a linear gradient (20%–60% acetonitrile in 0.1% TFA over 30 min). Identify Ub K48C (retention time is 17.2 min) and Ub-CAET-Acm (retention time is 16.2 min) peaks by mass spectrometry. Terminate the reaction when the product peak area plateaus (yield: 80%).

3.2.2 Purification of Ub-CAET-Acm (Steps 5–6)

1) Purify the reaction mixture by preparative HPLC (C18 column, same gradient as Step 4).
2) Collect the Ub-CAET-Acm peak fraction and lyophilize for 48 h. Store as a dry powder.

3.2.3 Removal of Acm Protecting Group (Steps 7–9)

1) Deprotection: Dissolve 16 mg of Ub-CAET-Acm in 0.9 mL deprotection buffer (6 M Gn·HCl, 0.2 M Na_2HPO_4, pH 7.5) to 2 mM. Add 60 μL of freshly prepared 0.5 M $PdCl_2$ solution (final: 30 mM) (*see* **Note 3**).
2) Reaction and Quenching: Incubate at 37°C (120 rpm) for 1 h. Terminate by adding 400 μL of 1 M DTT (*see* **Note 4**).
3) Purification: Isolate Ub-K48C-CAET via HPLC (same method as Step 5, retention time is 16.0 min, Fig. 2b, c). Lyophilize the product fraction.

3.2.4 Refolding of Ub K48C-CAET (Steps 10–15)

1) Dialysis: Dissolve 8 mg of lyophilized Ub-K48C-CAET in 1 mL refolding buffer. Transfer to 3 kDa MWCO dialysis tubing and dialyze sequentially (*see* **Note 5**):
2) 4 h in 1 L Dialysis Buffer 1 (1 M Gn·HCl, 0.2 M Na_2HPO_4, 1 mM TCEP, pH 7.5).
3) 4 h in 1 L Dialysis Buffer 2 (0.5 M Gn·HCl, 0.2 M Na_2HPO_4, 1 mM TCEP, pH 7.5).
4) Centrifugation: Clarify the dialysate by centrifugation (14,000 × *g*, 10 min). Discard insoluble precipitate.

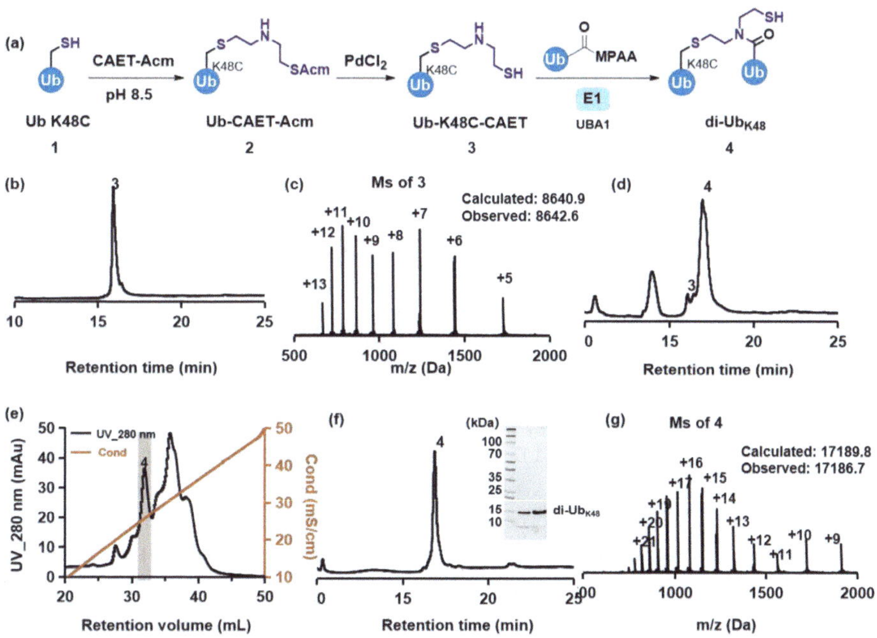

Fig. 2 Synthesis of K48-Linked Diubiquitin via Chemical Ligation. (**a**) Synthetic route for di-Ub_{K48} from Ub K48C. (**b**) Analytical RP-HPLC (214 nm) traces of Ub-K48C-CAET. (**c**) High-resolution ESI-MS spectra of Ub-K48C-CAET. (**d**) Analytical RP-HPLC (214 nm) traces for generation of di-Ub_{K48} (t = 3 h). (**e**) Purification of di-Ub_{K48} by cation-exchange chromatography. (**f**) Analytical RP-HPLC (214 nm) traces and SDS-PAGE of di-Ub_{K48}. (**g**) High-resolution ESI-MS spectra of di-Ub_{K48}

5) SEC Purification: Load the supernatant onto a Superdex 75 10/300 GL column pre-equilibrated with SEC buffer. Elute at 0.5 mL/min; Ub-K48C-CAET elutes at ~15 mL.

6) Concentration: Pool peak fractions and concentrate to 5 mg/mL.

3.2.5 Enzymatic Synthesis of CAET-Linked K48-Diubiquitin (Steps 16–17)

1) One-Pot Reaction: Combine 500 μL of 0.5 mM Ub (wild-type), 500 μL of 0.5 mM Ub-K48C-CAET, and 2 μM hUBA1 in reaction buffer (20 mM HEPES, 150 mM NaCl, 5 mM TCEP, 10 mM ATP, 10 mM $MgCl_2$, 25 mM MPAA, pH 7.5). Incubate at 37°C for 3 h. Monitor reaction completion by HPLC (yield: 80%, Fig. 2d) (*see* **Note 6**).

2) Purification: Isolate K48-diubiquitin using SOURCE™ 15 S cation-exchange chromatography (gradient: 0–500 mM NaCl in 50 mM sodium acetate, pH 4.5, the peak emergence conductivity was 24.18 mS/cm. Fig. 2e–g). Lyophilize the pure product.

3.3 Enzymatic Synthesis of Ubiquitinated Substrate

3.3.1 Peptide Synthesis

A type II degron substrate recognized by the E3 ligase UBR1 was utilized. The degron peptide (sequence: YIFSTDTGPGHLMK-EGYEEQLTE) with an SUE1 tag for enzymatic ubiquitination was synthesized by Genscript (*see* **Note 7**).

3.3.2 Enzymatic Reaction Setup

Combine 250 μL of 6.2 mg/mL di-Ub_{K48} (final concentration: 120 μ M) with 350 μL of 2 mg/mL degron peptide (final concentration: 300 μM) in reaction buffer (20 mM HEPES, 150 mM NaCl, 5 mM TCEP, 5 mM ATP, 10 mM $MgCl_2$, 25 mM MPAA, and pH 7.5). Initiate the reaction by adding 100 μL of 1 mg/mL UBA1 (final concentration: 2 μM) and 64 μL of 2.8 mg/mL UBE2E1 (final concentration: 2 μM). Incubate the mixture in a water bath at 37°C for 6 h.

3.3.3 Reaction Monitoring by HPLC-MS

Monitor the reaction using HPLC-MS. Take 50 μL aliquots from the reaction mixture, quench them with 900 μL of dilution buffer (6 M Gn·HCl, 0.2 M Na_2HPO_4, pH 8.0), and inject via a sample loop. Perform the analysis using an HPLC gradient from 20% to 60% mobile phase B over 30 min. The reaction provided over 95% conversion of di-Ub_{K48} (retention time is 16.3 min, Fig. 3b, c).

3.3.4 Purification

Purify the final reaction mixture using a Superdex 75 Increase 10/300 GL column. This procedure yields approximately 1 mg of pure diUb_{K48}-degron conjugate with an isolated yield of 70% (Fig. 3d).

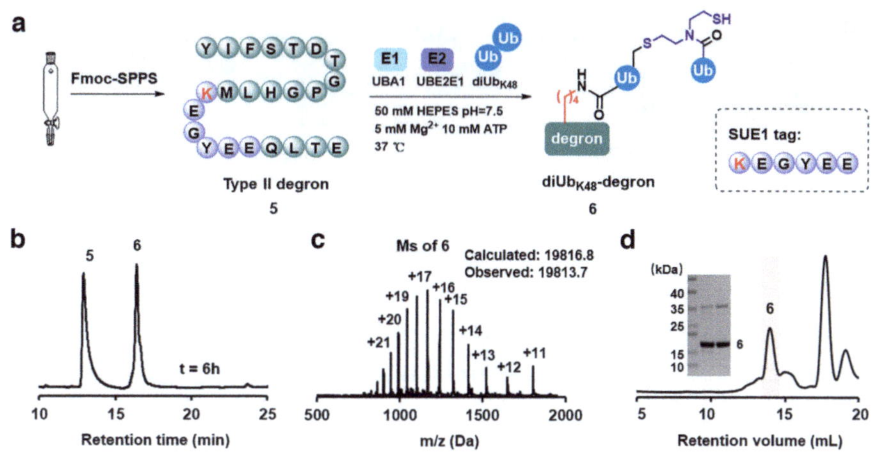

Fig. 3 Enzymatic Synthesis of Ubiquitinated Substrate. (**a**) Synthetic route for diUb $_{K48}$-degron. (**b**) Analytical RP-HPLC (214 nm) traces for generation of diUb $_{K48}$-degron ($t = 6$ h). (**c**) High-resolution ESI-MS spectra of diUb $_{K48}$-degron. (**d**) Purification of diUb $_{K48}$-degron by Superdex 75 Increase 10/300 GL column and SDS-PAGE of diUb $_{K48}$-degron

3.4 Construction of the K48-diUb Extension Probe

3.4.1 Activation of UBC2

Combine 1 mL of 150 μM UBC2 (3.0 mg/mL) with 10 equiv. DTNB in reaction buffer (20 mM HEPES, 150 mM NaCl, pH 7.5). Incubate the reaction at 37°C for 10 min to allow complete conversion of the starting material to the UBC2-TNB adduct.

3.4.2 Reaction Monitoring by HPLC-MS

Monitor the reaction progress by HPLC-MS. Dilute a 15 μL aliquot of the reaction mixture with 900 μL of dilution buffer (6 M Gn·HCl, 0.2 M Na_2HPO_4, pH 8.0) and inject via an auto-sampler. Perform the analysis using an analytical XB-C4 column with a gradient of 20%–60% mobile phase B over 30 min (retention time is 23.8 min, Fig. 4b). Confirm the formation of the UBC2-TNB product by mass spectrometry (Fig. 4c).

3.4.3 Purification of UBC2-TNB

Purify the reaction mixture using a Superdex 75 Increase 10/300 GL (SD75) column to yield approximately 2.5 mg of pure UBC2-TNB conjugate (Fig. 4d).

3.4.4 Assembly of the K48-diUb Extension Probe

Assemble the K48-diUb extension probe by combining 0.5 mg of the diUb$_{K48}$-degron conjugate with 2.5 mg (5 equiv.) of the purified UBC2-TNB in 500 μL reaction buffer (20 mM HEPES, 150 mM NaCl, pH 7.5). Incubate the mixture at 37°C for 30 min.

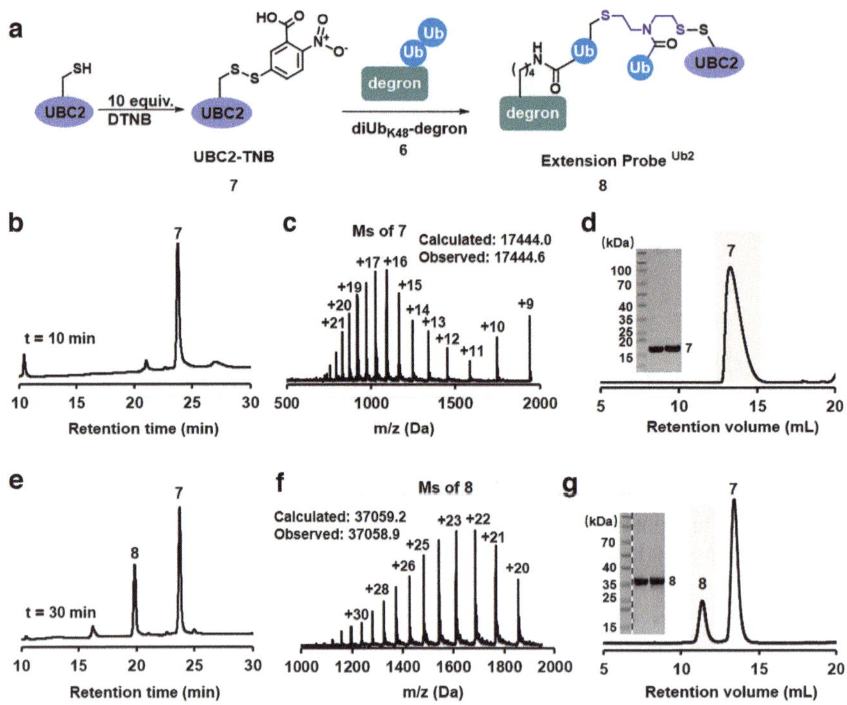

Fig. 4 Construction of the K48-diUb Extension Probe (Extension probe ^{Ub2}). (**a**) Synthetic route for K48-diUb Extension Probe. (**b**) Analytical RP-HPLC (214 nm) traces for generation of UBC2-TNB ($t = 10$ min). (**c**) High-resolution ESI-MS spectra of UBC2-TNB. (**d**) Purification of UBC2-TNB by Superdex 75 Increase 10/300 GL column and SDS-PAGE of UBC2-TNB. (**e**) Analytical RP-HPLC (214 nm) traces for generation of Extension probe ^{Ub2} ($t = 30$ min). (**f**) High-resolution ESI-MS spectra of Extension probe ^{Ub2}. (**g**) Purification of Extension probe ^{Ub2} by Superdex 75 Increase 10/300 GL column and SDS-PAGE of Extension probe ^{Ub2}

3.4.5 Analysis of Probe Assembly

Assess reaction completion by HPLC-MS. Dilute a 15 μL aliquot with 900 μL of dilution buffer and analyze using the same HPLC gradient and MS detection method described in Sect. 3.4.2 (retention time: 19.8 min, Fig. 4e). Verify the molecular weight of the final product by mass spectrometry (Fig. 4f) to confirm successful formation of the K48-diUb extension probe, with over 95% conversion of the $diUb_{K48}$-degron starting material.

3.4.6 Purification of the Ub Extension Probe

Subject the final reaction mixture to size-exclusion chromatography on a Superdex 75 Increase 10/300 GL column. This process yields approximately 0.4 mg of the pure ternary complex, designated as the K48-diUb extension probe (Fig. 4g) (*see* **Note 8**).

4 Notes

1. **Buffer Exchange**: Prior to enzymatic reactions, ubiquitin must be exchanged into a compatible HEPES-NaCl-based buffer.
2. **Conjugation pH**: Maintain the conjugation reaction pH at 8.5 and monitor it frequently to favor the reactive thiolate anion.
3. **Fresh $PdCl_2$**: The $PdCl_2$ solution for deprotection must be prepared fresh immediately before use.
4. **Quench Pd^{2}+:** After Acm deprotection, thoroughly quench the reaction with DTT to sequester Pd^{2}+ ions and prevent oxidation.
5. **Reducing Environment**: Include 1 mM TCEP in dialysis buffers to protect free thiols from oxidation and disulfide formation.
6. **Controlled Distal Unit Identity**: The engineered Ub K48C-CAET is designed to undergo self-cyclization upon MPAA activation, preventing its incorporation as the distal ubiquitin. This ensures that the resulting K48-diUb product exclusively contains wild-type Ub at the distal position.
7. **Recognition Tag**: The SUE1 tag (KEGYEE) on the substrate is essential for its recognition and ubiquitination by the E2 enzyme UBE2E1.
8. **Probe Stability**: Aliquot and flash-freeze the final Ub extension probe to preserve its activity, avoiding repeated freeze-thaw cycles.

Acknowledgments

This work was financially supported by the National Natural Science Foundation of China (Nos. 22034004 and 22027807), the New Cornerstone Science Foundation, and the Beijing Life Science Academy Initiative Scientific Research Program (No. 2023000CA0050).

References

1. Komander D, Rape M (2012) The ubiquitin code. Annu Rev Biochem 81(1):203–229
2. Dikic I, Schulman BA (2023) An expanded lexicon for the ubiquitin code. Nat Rev Mol Cell Biol 24(4):273–287
3. Yau R, Rape M (2016) The increasing complexity of the ubiquitin code. Nat Cell Biol 18(6):579–586
4. Swatek KN, Komander D (2016) Ubiquitin modifications. Cell Res 26(4):399–422
5. Buetow L, Huang DT (2016) Structural insights into the catalysis and regulation of E3 ubiquitin ligases. Nat Rev Mol Cell Biol 17(10):626–642
6. Genschik P, Sumara I, Lechner E (2013) The emerging family of CULLIN3-RING ubiquitin

ligases (CRL3s): cellular functions and disease implications. EMBO J 32(17):2307–2320
7. Popovic D, Vucic D, Dikic I (2014) Ubiquitination in disease pathogenesis and treatment. Nat Med 20(11):1242–1253
8. Rape M (2018) Ubiquitylation at the crossroads of development and disease. Nat Rev Mol Cell Biol 19(1):59–70
9. Kubbutat MH, Jones SN, Vousden KH (1997) Regulation of p53 stability by Mdm2. Nature 387(6630):299–303
10. Pao K-C, Stanley M, Han C et al. (2016) Probes of ubiquitin E3 ligases enable systematic dissection of parkin activation. Nat Chem Biol 12(5):324–331
11. Arkinson C, Walden H (2018) Parkin function in Parkinson's disease. Science 360(6386):267–268
12. Pickrell AM, Youle RJ (2015) The roles of PINK1, parkin, and mitochondrial fidelity in Parkinson's disease. Neuron 85(2):257–273
13. Békés M, Langley DR, Crews CM (2022) PROTAC targeted protein degraders: the past is prologue. Nat Rev Drug Discov 21(3):181–200
14. Teng X, Zhao X, Dai Y et al. (2024) ClickRNA-PROTAC for tumor-selective protein degradation and targeted cancer therapy. J Am Chem Soc 146(40):27382–27391
15. Wu T, Yoon H, Xiong Y et al. (2020) Targeted protein degradation as a powerful research tool in basic biology and drug target discovery. Nat Struct Mol Biol 27(7):605–614
16. Hehl LA, Horn-Ghetko D, Prabu JR et al. (2024) Structural snapshots along K48-linked ubiquitin chain formation by the HECT E3 UBR5. Nat Chem Biol 20(2):190–200
17. Horn-Ghetko D, Krist DT, Prabu JR et al. (2021) Ubiquitin ligation to F-box protein targets by SCF–RBR E3–E3 super-assembly. Nature 590(7847):671–676
18. Liang LJ, Chu GC, Qu Q et al. (2021) Chemical synthesis of activity-based E2-ubiquitin probes for the structural analysis of E3 ligase-catalyzed transthiolation. Angew Chem Int Ed Engl 60(31):17171–17177
19. Mabbitt PD, Loreto A, Déry MA et al. (2020) Structural basis for RING-Cys-relay E3 ligase activity and its role in axon integrity. Nat Chem Biol 16(11):1227–1236
20. Xia Q, Meng X, Wang Y et al. (2024) A cell-permeable Ub-Dha probe for profiling E1-E2-E3 enzymes in live cells. Chem Commun (Camb) 60(32):4342–4345
21. Li C, Wang T, Liang L et al. (2023) Simultaneous capture of ISG15 conjugating and deconjugating enzymes using a semi-synthetic ISG15-Dha probe. Sci China: Chem 66(3):837–844
22. Li C, Zhao F, Zuo C et al. (2025) Design and semisynthesis of ubiquitin extension probes for structural analysis of cullin1-mediated substrate polyubiquitination. J Am Chem Soc 147(27):23878–23890
23. Liang L-J, Wang Y, Hua X et al. (2023) Cell-permeable stimuli-responsive ubiquitin probe for time-resolved monitoring of substrate ubiquitination in live cells. JACS Au 3(10):2873–2882
24. Pan M, Zheng Q, Wang T et al. (2021) Structural insights into Ubr1-mediated N-degron polyubiquitination. Nature 600(7888):334–338
25. Ai H, Tong Z, Deng Z et al. (2025) Mechanism of nucleosomal H2A K13/15 monoubiquitination and adjacent dual monoubiquitination by RNF168. Nat Chem Biol 21(5):668–680
26. Ai H, Tong Z, Deng Z et al. (2023) Synthetic E2-Ub-nucleosome conjugates for studying nucleosome ubiquitination. Chem 9(5):1221–1240
27. Mao J, Ai H, Wu X et al (2023). Structural visualization of HECT-E3 Ufd4 accepting and transferring ubiquitin to form K29/K48-branched polyubiquitination on N-degron. BioRxiv: 2023.2005. 2023.542033
28. Tao S, He W, Li C et al. (2025) Synthesis of an E2-Ub-nucleosome conjugate to capture the E3 ligase PHF7-catalyzed H3K14 ubiquitination intermediate. Sci China: Chem 68:1–10.
29. Zhang L, Deng Z, Du Y et al. (2025) RAD18-catalysed formation of ubiquitination intermediate mimic of proliferating cell nuclear antigen PCNA. Bioorg Med Chem 117:118016
30. Baek K, Krist DT, Prabu JR et al. (2020) NEDD8 nucleates a multivalent cullin-RING-UBE2D ubiquitin ligation assembly. Nature 578(7795):461–466
31. Brown NG, VanderLinden R, Watson ER et al. (2016) Dual RING E3 architectures regulate multiubiquitination and ubiquitin chain elongation by APC/C. Cell 165(6):1440–1453
32. Liwocha J, Li J, Purser N et al. (2024) Mechanism of millisecond Lys48-linked poly-ubiquitin chain formation by cullin-RING ligases. Nat Struct Mol Biol 31(2):378–389
33. Deng Z, Ai H, Sun M et al. (2023) Mechanistic insights into nucleosomal H2B monoubiquitylation mediated by yeast Bre1-Rad6 and its human homolog RNF20/RNF40-hRAD6A. Mol Cell 83(17):3080–3094.e3014
34. Dawson PE, Muir TW, Clark-Lewis I et al. (1994) Synthesis of proteins by native chemical ligation. Science 266(5186):776–779
35. Han D, Cui Y, Deng X et al. (2025) Mechanically triggered protein desulfurization. J Am Chem Soc 147(5):4135–4146
36. Wang B, Li C, Zhao F et al. (2025) Examining the role of threonine phosphorylation in ubiquitin's function using chemical protein synthesis. JACS Au 5(5):2148–2158
37. Wu X, Du Y, Liang LJ et al. (2024) Structure-guided engineering enables E3 ligase-free and versatile protein ubiquitination via UBE2E1. Nat Commun 15(1):1266

Check for updates

Chapter 7

Biophysical and Cellular Assays for the Evaluation of Peptidomimetic Inhibitors Based on Degron Sequences

Christopher Lenz, Krishna Saxena and Stefan Knapp

Abstract

Scientific research in drug discovery relies on robust, versatile, and well-established methodologies. One effective strategy for targeting disease-relevant proteins, such as E3 ligases, involves rational drug design using degron peptides as starting points for compound development. Here, we present a comprehensive set of complementary assays for the evaluation of degron-based binders, enabling their integration into an efficient drug discovery pipeline. We introduce screening strategies such as Differential Scanning Fluorimetry (DSF) and Fluorescence Polarization (FP), alongside both solution-based and immobilization-based biophysical techniques, including Isothermal Titration Calorimetry (ITC) and Surface Plasmon Resonance (SPR) for reliable affinity determination. To assess intracellular interactions with full-length target proteins, we also employ Nano Bioluminescence Resonance Energy Transfer (NanoBRET). Together, these methods establish a robust framework for the discovery and characterization of degron-based peptidomimetic compounds.

Keywords DSF, SPR, FP, ITC, NanoBRET, Degron, Peptidomimetics, E3 ligase, PROTAC, K_D determination, Biophysical assays

1 Introduction

In the field of drug discovery, the exploration of endogenous mechanisms plays a crucial role in understanding disease-relevant processes. Recently, a novel approach called targeted protein degradation (TPD) has emerged as a promising strategy, aiming to degrade pathogenic proteins rather than merely inhibiting their biological activity [1]. PROteolysis TArgeting Chimeras (PROTACs) have emerged as one of the most important pharmacological modalities of TPD strategies. PROTACs have greatly expanded targeting capabilities, extending small molecule-based drug development via the ability to engage disease-causing proteins previously considered "undruggable," triggering their degradation by the ubiquitin-proteasome system (UPS) [2]. PROTACs are bifunctional molecules, consisting of an E3 ligase ligand, a linker moiety and a warhead that binds to the target, which

Maria W. Górna et al. (eds.), *Degrons*, Methods in Molecular Biology, vol. 3069,
https://doi.org/10.1007/978-1-0716-5508-5_7,

is then recognized as a neo-substrate by the UPS, resulting in selective target polyubiquitination and subsequent degradation [3]. An important step in the development of innovative PROTACs is the discovery of small molecules that tightly bind to E3 ligases [4].

Degron sequences act similarly to PROTACs, as they regulate the degradation of certain proteins by targeting them to the UPS. These short motifs, however, are either an inherent feature of protein sequences or they can be acquired by post-translational modifications [5]. Degrons typically interact directly with E3 ligases, making them ideal candidates for ligase-recruiting elements in PROTAC design. By mimicking the sequence of a degron and synthesizing peptidomimetics based on this motif, attachment points for E3 ligases can be developed [6–8].

For example, the E3 ligase Tripartite Motif Containing 7 (TRIM7) recognizes proteins with a C-terminal glutamine preceded by a hydrophobic residue via its C-terminal PRYSPRY domain, subsequently targeting them for proteolytic degradation (Fig. 1). In this context, the following protocols describe five distinct methods and evaluation strategies for the assessment and development of peptidomimetic ligands, featuring exemplary data derived from strategic approaches to target TRIM7 [6].

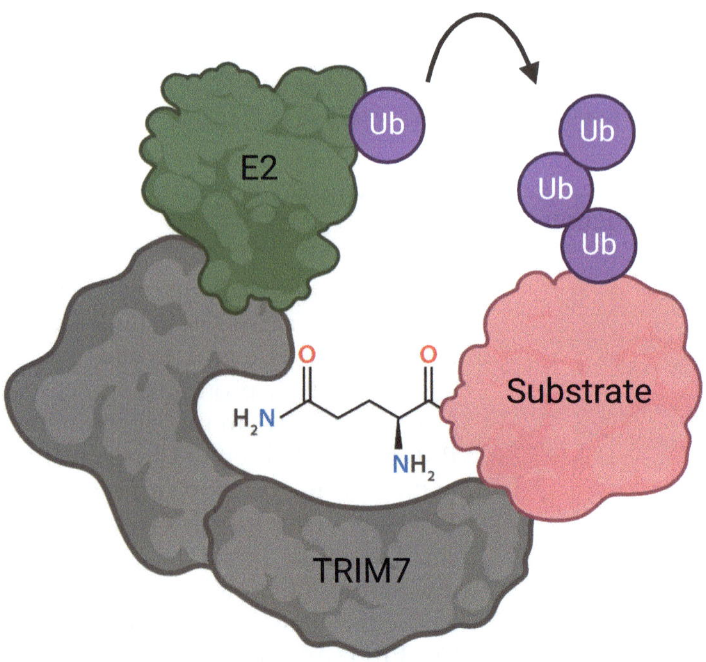

Fig. 1 Schematic illustration of TRIM7- and E2-mediated polyubiquitination of a substrate bearing a recognized glutamine residue

2 Materials

In the following sections, a purified E3 ligase protein is required for all *in vitro* protocols. Therefore, it is recommended to either purchase the protein commercially or express and purify it yourself. As a reference, Muñoz Sosa and Lenz et al. [6] provided an exemplary description of E3 ligase domain expression and purification.

2.1 Differential Scanning Fluorimetry (DSF)

1. DSF buffer: 20 mM HEPES, pH 7.5, 150 mM NaCl, 1 mM TCEP, 5× SYPRO Orange.
2. Dimethyl sulfoxide (DMSO).
3. Recombinant E3 ligase protein.
4. Peptidomimetic compounds/peptides.
5. 384-Well PCR Plate.
6. Serological pipettes.
7. Pipette controller.
8. MCE membrane filter, 0.45 μm pore size.
9. Polyolefin sealing film.
10. Real-Time PCR instrument.
11. Automated liquid dispenser.
12. 1.5 mL and 2 mL tubes.

2.2 Fluorescence Polarization (FP)

1. FP buffer: 50 mM HEPES, pH 7.4, 150 mM NaCl, 1 mM TCEP, 5% Glycerol, 0.05% Tween20.
2. Dimethyl sulfoxide (DMSO).
3. Cyanine5 (Cy5)-labeled peptide.
4. Recombinant E3 ligase protein.
5. Peptidomimetic compounds/peptides.
6. Black 384-well flat-bottom plate.
7. Serological pipettes.
8. Pipette controller.
9. MCE membrane filter, 0.45 μm pore size.
10. Microplate reader.
11. Transcreener-specific FP module (FP 590 675 675).
12. Automated liquid dispenser.
13. 1.5 mL and 2 mL tubes.

2.3 Surface Plasmon Resonance (SPR)

1. Running buffer: 10 mM HEPES, pH7.5, 150 mM NaCl, 1 mM TCEP, 0.05% Tween20.
2. Protein dilution buffers:

   a. 10 mM acetate, pH 4.5.
   b. 10 mM acetate, pH 5.0.
   c. 10 mM acetate, pH 5.5.
   d. 10 mM Bis-Tris, pH 6.0.
   e. 10 mM Bis-Tris, pH 6.5.

3. Recombinant E3 ligase protein.
4. Peptidomimetic compounds/peptides.
5. Dimethyl sulfoxide (DMSO).
6. 483 mM N-ethyl-N′-(3-(dimethylamino)propyl)carbodiimide (EDC).
7. 10 mM N-hydroxysuccinimide (NHS).
8. 1 M ethanolamine.
9. 1 M NaCl in 50 mM NaOH.
10. PP microplate, 96-well, U-shape.
11. 25 mL polystyrene reservoirs.
12. Serological pipettes.
13. Pipette controller.
14. MCE membrane filter, 0.2 μm pore size.
15. SPR instrument.
16. Sensor chip CM5.
17. Sensor chip NA.
18. Sensor chip SA.
19. 1.5 mL and 2 mL tubes.
20. 15 mL tubes.
21. 7 mm polypropylene plastic vials.
22. Rubber caps, type 3.
23. 15 mm polypropylene plastic vials.
24. Rubber caps, type 5.

2.4 Isothermal Titration Calorimetry (ITC)

1. ITC Buffer: 20 mM HEPES, pH 7.5, 150 mM NaCl, 1 mM TCEP.
2. Dimethyl sulfoxide (DMSO).
3. Contrad 70.
4. Recombinant E3 ligase protein.
5. Peptidomimetic compounds/peptides.
6. Serological pipettes.
7. Pipette controller.
8. MCE membrane filter, 0.45 μm pore size.

9. 15 mL ultra-centrifugal filter, 10 kDa cutoff.
10. ITC instrument.
11. 1.5 mL and 2 mL tubes.
12. 15 mL tubes.

2.5 Nano Bioluminescence Resonance Energy Transfer (NanoBRET)

1. Dulbecco's modified Eagle's medium (DMEM), 10% fetal bovine serum (FBS), 1% penicillin-streptomycin (PS).
2. Opti-MEM I reduced serum medium.
3. Dulbecco's phosphate-buffered saline (DPBS).
4. 0.05% Trypsin-EDTA.
5. Dimethyl sulfoxide (DMSO).
6. Eukaryotic expression vector with full-length protein fused to Nluc (Nluc-E3-ligase DNA) [9].
7. Transfection Carrier DNA.
8. BODIPY-labeled peptide/peptidomimetic.
9. Peptidomimetic compounds/peptides.
10. HEK293T cells or a different appropriate cell line.
11. Nano-Glo Substrate (Promega).
12. FuGENE HD Transfection Reagent (Promega).
13. White 384-well flat-bottom plate.
14. Microplate reader with a luminescence filter pair (450 nm BP filter and 610 nm LP filter).
15. Automated liquid dispenser.
16. Serological pipettes.
17. Pipette controller.
18. Cell culture flask T75.
19. 1.5 mL and 2 mL tubes.
20. 15 mL tubes.
21. Automated cell counter.
22. CO_2 incubator.
23. Laminar flow hood.

3 Methods

In general, the use of an automated liquid dispenser, such as the Echo 550 acoustic dispenser (Labcyte), is highly recommended to allow a fast throughput of measurements in a significantly reduced assay volume, resulting in much lower cost per data point.

3.1 Differential Scanning Fluorimetry

Differential Scanning Fluorimetry (DSF) is an easy-to-use but broadly applicable high-throughput in vitro screening method to determine and analyze the apparent melting temperature (Tm) of a purified protein in bound and unbound states [10]. As the temperature increases, the protein unfolds, exposing its hydrophobic regions, which can be bound by a dye such as SYPRO Orange. This interaction causes a substantial increase in fluorescence, serving as an indicator of the protein unfolding transition (*see* **Note 1**) [11].

Samples used in DSF are prepared as triplicates to determine the standard deviation for each compound analyzed. A known binder should be added as a positive control (e.g., peptide degron sequence) as well as a negative/no compound control with only the protein and solvent present to reference data and determine the relative ΔTm. Moreover, an additional protein-free control for auto-fluorescent compounds or buffer ingredients is helpful to prevent the misinterpretation of artifactual, protein-independent fluorescence. For the DSF measurements conducted in this work, we used a QuantStudio 5 Real-Time PCR machine, the Applied Biosystems QuantStudio Design & Analysis Software, and Applied Biosystems Protein Thermal Shift Software.

3.1.1 DSF Sample Preparation and Data Generation

1. Dilute the E3 ligase protein to 10 μM in 0.45 μm filtered DSF Buffer (*see* **Note 2**).
2. Transfer 10 μL of diluted protein to each well of a 384-well PCR plate.
3. Add peptides/peptidomimetic compounds to reach 100 μM (10× protein concentration) (*see* **Note 3**) with 1%–5% DMSO using an automated liquid dispenser to three separate wells for triplicates; additionally, leave a triplicate as an “only protein” control (*see* **Note 4**).
4. Seal the plate using a polyolefin sealing film and place it into the real-time PCR instrument.
5. Set up the experiment method by setting the cover temperature to 105°C, the starting temperature to 25°C, the temperature increase rate to 0.05°C/s, and the final temperature to 85°C.
6. Select respective wells with your samples and define ligand (protein) and analyte (peptides/peptidomimetics).
7. Start the run.
8. After ~25 min, evaluate the data using a protein thermal shift analysis software.

3.1.2 DSF Data Evaluation

For the evaluation of your data, it is important to note that there are two most commonly used approaches.

1. ΔTm-Derivative analysis.
2. ΔTm-Boltzmann fit.

Both evaluation methods can be used to determine Tm values; however, the strategy used for assessment of the data should stay the same.

For the derivative analysis, the derivative of the fluorescence signal with respect to the temperature (d(Fluorescence)/dT) (Fig. 2B) is being determined. Here, the Tm corresponds to the maximum reached for the derivative.

The Boltzmann fit (Fig. 2A) applies a sigmoidal model to fit the melting curve as a function of temperature. Respective Tm values are determined as the temperature at which 50% of the protein is unfolded, corresponding to the inflection point of the curve.

By providing an "only protein" reference, ΔTm values can be determined via:

$$\Delta Tm = Tm(protein\ with\ compound) - Tm(protein\ without\ compound)$$

3.2 Fluorescence Polarization

The fluorescence polarization (FP) assay is an advanced technique to monitor molecular interactions in solution and is based on the principle that small fluorescently labeled molecules (typically <1.5 kDa) exhibit slower rotational motion when bound to a larger partner (typically >10 kDa). Polarization values, calculated as the ratio of the difference between vertically and horizontally polarized light to their sum, serve as indicators of the ligand state, with low polarization values corresponding to the unbound state and high values indicating the bound state [12].

This method is suitable for screening molecules, such as peptidomimetic compounds, against E3 ligase targets, enabling an initial assessment of ligand affinity. The FP assay format also serves to confirm the binding results of investigated compounds determined by DSF. In general, Tm values correlate with binding affinities observed in an FP competition-based approach, but the magnitude of ΔTm

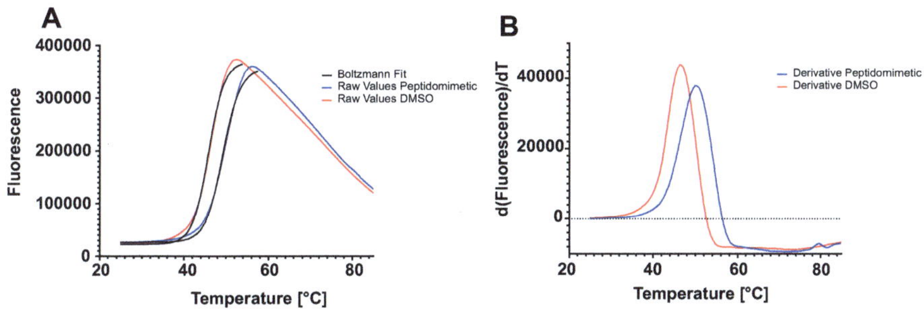

Fig. 2 (**A**, left) Exemplary thermal shift melting curves of TRIM7^{PRYSPRY} with DMSO (blue) and TRIM7^{PRYSPRY} with a peptidomimetic compound (red) as well as the overlaid Boltzmann fit (black). (**B**, right) Derivative curves of the same data as in (**A**)

shifts may depend on the E3 ligase used. By conjugating a fluorescent dye, such as Cy5, to the relevant degron peptide sequence, a tracer molecule can be easily generated (*see* **Note 5**). Prior to its use in a competition assay format, the tracer molecule needs to be analyzed via tracer titration with the protein (E3 ligase) to determine its effective affinity.

All samples are prepared as triplicates; additionally, if the assay is performed for the first time, different tracer concentrations should be used to determine optimal polarization intensities. For the following FP measurements, we used a Pherastar FSX plate reader.

3.2.1 FP Tracer Titration Assay

1. Dilute the tracer in 0.45 μm filtered FP buffer to reach different tracer concentrations (e.g., 0.1–0.3× expected K_D of the tracer) (*see* **Note 6**).
2. Add up to e.g. 100 μM E3 ligase to 100 μL of each tracer suspension.
3. Serially dilute the protein-tracer suspension using a 1:1 ratio by following these steps: Sequentially add 50 μL of tracer-only suspension to 50 μL of the preceding E3-ligase-tracer suspension. Repeat this process for a total of 10–15 dilution steps.
4. Transfer 3 × 10 μL of each dilution to three separate wells of a black 384-well flat-bottom plate.
5. 3 × 10 μL of the initial tracer suspension in FP buffer with no protein should also be added to the plate to set the baseline (unbound tracer sample).
6. Incubate the plate at room temperature (RT) in dark conditions for 40 min.
7. Read the perpendicular and parallel intensities using a microplate reader.
8. Calculate the mean polarization values [mP] for each concentration:

$$Polarization\ [mP] = \frac{parallel - perpendicular}{parallel + perpendicular} \times 1000$$

9. Calculate the baseline-corrected polarization (P) by subtracting the polarization value of the unbound tracer sample from all other sample values. Plot the corrected polarization values P against the corresponding tracer concentrations (C) (Fig. 3A). Perform a nonlinear regression fit to determine the K_D value of the tracer molecule:

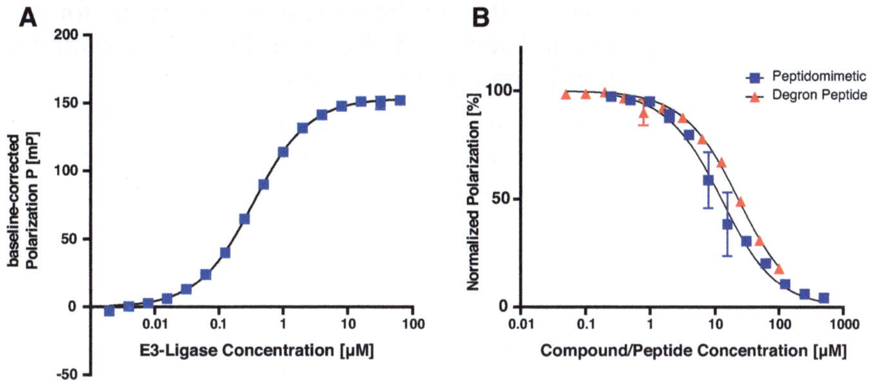

Fig. 3 (**A**, left) Exemplary FP tracer titration curve for the interaction of E3 ligase TRIM7^{PRYSPRY} with a peptidomimetic tracer. (**B**, right) Dose–response competition curves of a peptidomimetic compound (blue) and the degron peptide (red) against TRIM7^{PRYSPRY}. Non-linear regression fits are overlaid in black

$$P\,[mP] = \frac{P_{max} \times C}{K_D + C}$$

where P is the polarization at a given concentration, P_{max} represents the maximum polarization, C is the concentration of the tracer, and K_D is the dissociation constant of the tracer.

3.2.2 FP Dose–Response Competition Assay

1. Prepare the tracer solution by diluting the tracer in FP buffer to a final concentration of 0.1–0.3× K_D of the tracer–protein interaction, as determined previously (see Sect. 3.2.1).
2. Before adding the protein to the tracer solution, take out 100 μL to have an unbound tracer sample.
3. Add protein at a concentration of 2–3× K_D to generate the protein-tracer mixture, then take out 100 μL for the 100% bound tracer sample.
4. Transfer 10 μL to each well of a black 384-well flat-bottom plate.
5. Use an automated liquid dispenser to titrate different concentrations of peptidomimetic compounds at 1%–5% final DMSO concentration to each well of the protein-tracer mixture (*see* **Note** 7).

Alternatively, if you don't have an automated liquid dispenser:

1. Prepare the tracer solutions diluted in FP buffer as described before (see Sect. 3.2.2, Steps 1 and 2).

2. Set aside 100 μL of the protein-tracer mixture for each compound and add the desired final DMSO concentration (1%–5%) to the remaining protein-tracer solution.
3. Add the highest concentration of compound to the 100 μL protein-tracer mixture set aside, and adjust the final DMSO concentration to the desired level.
4. Serially dilute the compound solution from Step 3 by transferring 50 μL of this mixture to 50 μL of the solution without compound.
5. Transfer 10 μL of each concentration into the corresponding wells of a black 384-well flat-bottom plate.
6. Incubate the plate for 40 min at RT under dark conditions.
7. Measure the perpendicular and parallel intensities using a microplate reader and calculate the mean polarization values (see Sect. 3.2.1, Step 8).
8. Normalize mean polarization values for each concentration:

$$\begin{aligned} &\mathit{Normalized\ Polarization}(Y)\ [\%] \\ &= \frac{\mathit{Polarization}(Y) - \mathit{Polarization}(\mathit{Unbound})}{\mathit{Polarization}(\mathit{Bound}) - \mathit{Polarization}(\mathit{Unbound})} \times 100 \end{aligned}$$

9. Plot the normalized polarization values (Y) versus concentration (C) and perform a nonlinear regression fit to determine the IC_{50} for each peptidomimetic compound (Fig. 3B):

$$Y = \frac{100}{1 + \frac{C}{IC_{50}}}$$

10. To determine K_i values for respective compounds using the FP competition assay, the following equation published by Nikolovska-Coleska et al. [13] is recommended:

$$K_i = \frac{[I]_{50}}{\frac{[L]_{50}}{K_D} + \frac{[P]_0}{K_D} + 1}$$

$[I]_{50}$ represents the concentration of free compound at 50% binding, $[L]_{50}$ denotes the free tracer concentration at 50% binding, $[P]_0$ corresponds to the free protein concentration at 0% binding, and K_D refers to the previously determined dissociation constant of the protein-tracer interaction. For more details, please refer to the publication by Nikolovska-Coleska et al.

3.3 Surface Plasmon Resonance

Surface Plasmon Resonance (SPR) acts as a label-free, non-fluorescent, and versatile analytical technique with high sensitivity, widely used for studying molecular interactions in real time. The principle of SPR depends on the interaction of polarized light with a thin metal film, typically gold, under conditions of total internal reflection [14]. When light hits the metal surface at a specific angle, it excites collective oscillations of conduction electrons, known as surface plasmons [15]. These oscillations produce evanescent waves that penetrate a short distance into the adjacent medium. The intensity of these waves is influenced by the refractive index near the chip's surface. At a specific angle, maximum energy transfer occurs, leading to a noticeable dip in the reflected light intensity [14, 16].

The resonance angle is highly sensitive to changes in the refractive index, caused by interactions on the metal surface. Ligands, such as E3 ligases, can be immobilized on the sensor surface, allowing the quantitative detection of binding events with analytes, such as peptidomimetic compounds. The magnitude of the refractive index change directly correlates with the mass and density of the interacting molecules, enabling the determination of kinetic parameters, such as association (k_a) and dissociation rates (k_d), as well as the equilibrium binding constant (K_D) [14, 16]. To validate potential binders identified in previous screening assays, SPR is an ideal method to accurately determine their respective affinities in a high-throughput manner.

The following protocol is specified for running assays on a Biacore T200 instrument with the Biacore T200 Control/Evaluation Software, using the PRYSPRY domain of the E3 ligase TRIM7, but it is broadly applicable to any other SPR instrument or protein-ligand system. There is a wide variety of immobilization strategies to robustly attach a protein to the surface of a sensor chip; however, the two most common approaches, amine coupling on CM5-Chips and biotin-neutravidin (NA)/-streptavidin (SA) coupling, will be described.

3.3.1 Preparation Steps for Immobilization of the Protein (See Note 8)

1. Prepare 1.5 L running buffer using deionized water and filter with a 0.2 μm pore size membrane filter.
2. Take out 500 mL as immobilization buffer and 50 mL running buffer for dilution purposes (*see* **Note 9**).
3. Attach the immobilization buffer to the SPR machine and prime the system.
4. Eject the previous sensor chip from the instrument and inject a newly opened sensor chip CM5 or NA-/SA-Chip, then prime again.
5. Add 19.4 mL of 100% DMSO to 950 mL running buffer to reach 2% DMSO (for 5% DMSO, add 50 mL).

3.3.2 Scouting for an Optimal pH Level of the Protein Dilution Buffer (for CM5 Chips)

1. Start a manual run at a 10 μL/min flow rate on any flow channel (FC) except for FC1.
2. Dilute the protein to 10 μg/mL separately in protein dilution buffers a-e, with pH values ranging from 4.5 to 6.5.
3. Inject each protein dilution for 30 s and monitor the binding to the chip surface.
4. Select the most suitable buffer by choosing the highest pH, which is still showing a high response increase for further covalent immobilization procedure (*see* **Note 10**).

3.3.3 Protein Immobilization on Sensor Chip CM5

1. Prepare 100 μL of 483 mM N-ethyl-N′-(3-(dimethylamino)propyl)carbodiimide (EDC), 100 μL of 10 mM N-hydroxysuccinimide (NHS), and 140 μL of 1 M ethanolamine separately in a 7 mm plastic vial.
2. Add up to 700 μL of 10 μg/mL E3 ligase diluted in protein dilution buffer to another 7 mm plastic vial and keep an additional 7 mm plastic vial empty for the mixture of EDC and NHS.
3. Select a new immobilization method and choose “Amine coupling” on the second FC with the following settings:
   i. Mix and inject 483 mM EDC + 10 mM NHS (1:1) for 420 s at 10 μL/min.
   ii. Inject 10 μg/mL protein diluted in protein dilution buffer at 10 μL/min with a contact time according to the desired immobilization level (*see* **Note 11**).
   iii. Inject 1 M Ethanolamine for 420 s at 10 μL/min.
4. Start the immobilization run and repeat Steps 1–3 for the remaining third and fourth FCs to generate replicate protein surfaces.
5. Repeat Steps 1–3 additionally for the reference FC1, but select “Blank immobilization,” leaving out the protein injection.

3.3.4 Protein Immobilization on a Sensor Chip NA/SA

1. Start a manual run and select all FCs at 10 μL/min.
2. Consecutively inject 1 M NaCl in 50 mM NaOH for 1 min ×3 over all four FCs.
3. End the run and start a new manual run by selecting FC2 at 10 μL/min.
4. Inject 10 μg/mL biotinylated E3 ligase diluted in immobilization buffer until the desired protein immobilization level is reached (*see* **Note 11**).
5. Stop the injection and check if the protein level stays stable.
6. Stop the run and repeat Steps 3–5 for FC3 and FC4 to create replicates.

3.3.5 Analysis of the Dose–response Binding of Peptidomimetic Compounds

1. Design a new method by setting the detection to all four channels, with FC1 being the reference channel and the sample compartment temperature set to 25°C.
2. Multi-cycle kinetics should be used as the experimental strategy with a flow rate of 30 μL/min as well as at least 60 s contact time and 150 s dissociation time.
3. Add the following assay steps while selecting ten compound concentrations, typically ranging from 0.1× to 10× the expected K_D for each analyte and include two blanks (running buffer without compound) per analyte. Ten startup injections should be added to ensure a stable baseline, while control sample injections are used to confirm protein integrity over time, using the peptide as a positive control (Fig. 4).
4. For a 2% DMSO solvent correction, prepare 10 mL 1% DMSO running buffer and 10 mL 3% DMSO running buffer, using the running buffer that was set aside in Sect. 3.3.1, Step 2 (*see* **Note 12**).
5. Mix both solutions using the pipetting scheme in Table 1 and transfer each of the eight mixtures to 7 mm plastic vials.
6. Prepare 300 μL of the highest compound concentrations at 2% DMSO using the running buffer that was set aside in Sect. 3.3.1, Step 2.
7. Transfer the entire volume to a designated well of a 96-well microplate, along with 150 μL of 2% DMSO running buffer into 11 adjacent wells for each compound.
8. Prepare a 1:1 serial dilution by transferring 150 μL from the highest concentration well to the adjacent well and repeat the process nine times sequentially, leaving two samples as blanks.
9. Prepare the control sample with a peptide concentration equal to its K_D and transfer the sample to a 7 mm plastic vial.

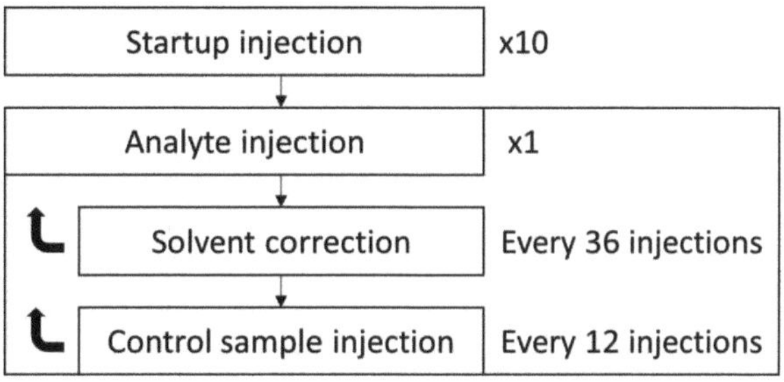

Fig. 4 Schematic representation of a general SPR workflow, starting with 10 startup injections, followed by repeated analyte injections, periodically followed by solvent corrections and control sample injections

Table 1
Pipetting scheme for the preparation of the SPR solvent correction

Mixture	1	2	3	4	5	6	7	8
1% DMSO running buffer	400 µL	350 µL	300 µL	250 µL	200 µL	150 µL	100 µL	50 µL
3% DMSO running buffer	100 µL	150 µL	200 µL	250 µL	300 µL	350 µL	400 µL	450 µl

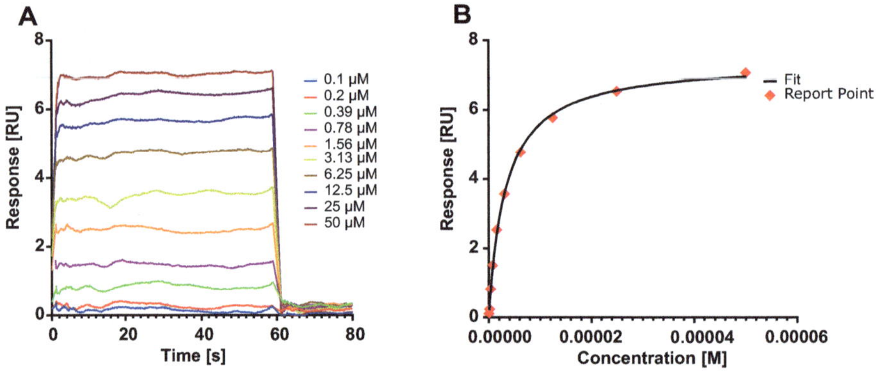

Fig. 5 (**A**, left) Exemplary SPR sensorgram overlay plot of a peptidomimetic compound interacting with $TRIM7^{PRYSPRY}$. (**B**, right) Steady-state affinity analysis of the same data as in (**A**) with respective report points (red) and overlaid steady-state affinity fit (black)

10. Transfer all necessary solutions as well as the 96-well microplate to the SPR metal rack and insert the rack into the sample compartment.
11. Attach the running buffer with 2% DMSO to the SPR instrument, prime the system, and eventually start the run.
12. After the run has finished, assess the data and plot the resulting Sensorgrams (Fig. 5A) using the SPR Evaluation Software.
13. If steady state was reached for all concentrations, evaluation can be done using the steady state affinity model (Fig. 5B) or manually via GraphPad Prism with the following equation:

$$R_{eq} = \frac{C \times R_{max}}{K_D + C} + offset$$

C stands for each concentration of the respective analyte, R_{max} is the maximum response, R_{eq} represents the response at steady state, and offset is the response at zero compound concentration.

14. For kinetic evaluation of the data (*see* **Note 13**), the 1:1 Langmuir interaction fit can be applied to determine the association rate constant k_a and dissociation rate constant k_d. The affinity can be calculated using the following equation:

$$K_D = \frac{k_d}{k_a}$$

3.4 Isothermal Titration Calorimetry

Isothermal Titration Calorimetry (ITC) is another widely accepted and powerful technique in the field of biophysical drug discovery, providing critical insights into molecular interactions. It serves as a complementary method to SPR, as ITC does not rely on binding kinetics but instead determines the thermodynamic profile of interactions [17]. While the throughput of this method is quite limited because of its substantial reagent requirements, ITC can be used as a versatile alternative technique to confirm potential binders. By directly measuring the heat released or absorbed during a binding event, ITC enables the determination of key binding parameters, including the binding affinity (K_D), enthalpy change (ΔH), entropy change (ΔS), and stoichiometry (n) of the interaction [18].

Thermodynamic parameters are calculated using the following equations, while the stoichiometry is derived from the molar ratio of the binding partners in the titration curve:

$$RT\ln(K_D) = \Delta G = \Delta H - T\Delta S$$

Here, R represents the gas constant, T is the temperature, and ΔG is the Gibbs free energy change.

The principle of ITC is based on successive titrations of an analyte (e.g., peptidomimetics) into a sample cell containing its binding partner (e.g., E3 ligase), while the whole process is performed under controlled isothermal and isobaric conditions [19]. Each binding event leads to a detectable heat change, which is recorded and analyzed to generate a thermodynamic binding profile [20]. Unlike most other biophysical techniques, ITC requires no immobilization or modification of interacting molecules, making it a label-free and solution-based method that consistently maintains the native state of both binding partners [21].

The following protocol is specifically optimized to run ITC for the PRYSPRY domain of E3 ligase TRIM7 on a TA Instruments Nano ITC device. However, it can be applied to any other E3 ligase/protein and ITC instrument by adjusting variable parameters like stirring speed, temperature, buffer conditions, the number of injections, analyte, and protein concentrations, as well as the DMSO concentration.

3.4.1 Sample Preparation

1. Given the high level of sensitivity in ITC, the composition of all components of the ITC buffer needs to be identical.
2. If necessary, exchange buffer for the protein, using a suitable centrifugal filter with a 10 kDa pore size at 4000 g, adding freshly prepared 0.45 μm filtered ITC buffer at least four times.
3. The protein sample (titrand) should have a final concentration of 30 μM in 1.6–1.8 mL ITC buffer with 0.6% DMSO.
4. Add 1.8 μL of 50 mM peptidomimetic compound in 100% DMSO to 298.2 μL ITC buffer for a 300 μM titrant compound solution.

3.4.2 ITC Titration Experiment

1. Prior to compound titration, all samples need to be degassed to prevent the formation of bubbles, which could affect thermal signals.
2. Simultaneously, clean the sample cell with detergent (e.g., Contrad 70), followed by rinsing with at least 300 mL deionized water, ensuring the cell is left empty.
3. Overfill the sample cell slightly above the desired level with an appropriate amount of titrant, then add 250 μL of the titrant to the syringe, ensuring no bubbles are present.
4. Place the syringe in the sample cell and start stirring at 300 rpm to equilibrate the system until the curve has reached steady state.
5. The titration is set to begin with a 4 μL injection of the titrant, followed by 27 injections of 8 μL at 25°C.

3.4.3 ITC Data Analysis

1. Analysis is performed by reviewing the thermogram, plotting raw heat rate (μcal/s) versus time (s), and checking for inconsistencies like baseline drifts, noise, or injection artifacts (Fig. 6 top panel).
2. Integrate the area under each peak to obtain the heat change (ΔQ) for every injection.
3. Exclude injection outliers and subtract the baseline ΔQ to remove any noise or systematic deviations.
4. Normalize the heat values by dividing the baseline-corrected heat change per injection by the number of moles of injected compound.
5. Plot the normalized heat (kcal/mol of injection) against the molar ratio of compound and protein to generate a binding isotherm and confirm that the data follow a sigmoidal curve, indicating saturation of the absolute heat change (Fig. 6 lower panel).

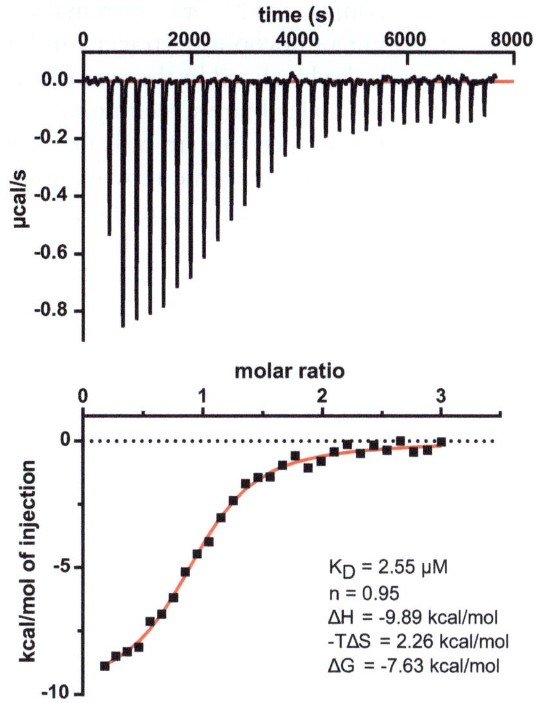

Fig. 6 ITC data of a peptidomimetic compound titrated to TRIM7^{PRYSPRY} depicted as a thermogram (top) and integrated heat data (bottom) fitted with a one-site binding model (red). Exemplary results for thermodynamic parameters are shown in the lower panel

6. Select an appropriate fitting model; in this instance, a one-site (1:1) binding model is assumed for a single binding site with the following equation:

$$Q = \frac{n[M]_t V_0 \Delta H}{1 + \frac{K_D}{[L]_t}}$$

Here, $[M]_t$ stands for the total protein concentration, V_0 represents the total volume of the ITC sample cell, and $[L]_t$ is the total compound concentration in solution.

3.5 Nano Bioluminescence Resonance Energy Transfer

Cellular assays play a crucial role in confirming potential binders in vivo. Here, the focus lies on the permeability of the compounds and their engagement with the full-length target in a cellular environment that is difficult to determine with most conventional in vitro assays. Nano bioluminescence resonance energy transfer (NanoBRET) is a powerful method that uses proximity-induced energy transfer of a bioluminescent enzyme to fluorescent acceptor molecules to measure this interaction upon substrate consumption [22]. In this context, a NanoLuc luciferase (Nluc) acts as the respective enzyme, conjugated N- or C-terminally to an E3 ligase (e.g., TRIM7), exciting a fluorophore (tracer) in the vicinity. The tracer molecule needs to be localized in the area of the potential peptidomimetic binding site to facilitate screening in a competitive manner [23]. Transiently transfected human cells like HEK293T are typically used to study cellular engagement, while the tracer molecule can be based on the peptide degron sequence or a peptidomimetic conjugated to the fluorescent dye BODIPY.

The following workflow describes a general NanoBRET assay process used to determine the apparent K_D of a tracer molecule to its protein target as well as the IC_{50} of compounds displacing the tracer. Both assay options require transient transfection of the Nluc-tagged E3 ligase before measurement and can be evaluated in intact or permeabilized cellular states. All samples should be prepared as triplicates to ensure reproducibility. Additionally, compounds, tracer, DMSO, and digitonin should be dispensed using an automated liquid dispenser to enable rapid and precise distribution. We used a Pherastar FSX plate reader for the following NanoBRET measurements.

3.5.1 Transfection of Nluc-Constructs

1. Prepare 500 µL transfection mix by adding 15 µL FuGENE HD transfection reagent to a mixture of 9 ng/µL Transfection Carrier DNA and 1 ng/µL Nluc-E3-ligase DNA construct in Opti-MEM medium (For the selection and design of functional Nluc-constructs, refer to Schwalm et al. [9]).
2. Incubate the transfection mix for 15 min at RT.
3. Seed HEK293T cells to a final density of 2.0×10^5 cells/mL in DMEM with 10% FBS and 1% PS, then add 500 µL transfection mix to 10 mL of cell solution in a T75 cell culture flask.
4. Incubate cells overnight at 37°C and 5% CO_2.

3.5.2 NanoBRET Tracer Titration Assay

1. Remove DMEM culture medium and displace with Opti-MEM medium using serological pipettes (*see* **Note 14**).
2. Adjust cell density to 2.0×10^5 cells/mL and transfer 10 µL of transfected cells to each well of a white 384-well flat-bottom plate.
3. Titrate different concentrations of the tracer into respective wells with a final DMSO concentration of a maximum of 1% (*see* **Note 15**).
4. Incubate the plate at 37°C and 5% CO_2 for 2 h.
5. Add 5 µL Nano-Glo substrate diluted 1:200 in Opti-MEM to each well.
6. Measure BRET ratio using a microplate reader with a luminescence filter pair [450 nm BP filter and 610 nm LP filter].
7. Lyse cells by adding 25 nL of 20 mg/mL digitonin to each well, then incubate for 5 min.

8. Measure the BRET ratio again with the same settings as in Step 6 for permeabilized cells.

3.5.3 NanoBRET Tracer Displacement Assay

1. Remove DMEM culture medium and displace with Opti-MEM medium using serological pipettes.
2. Adjust cell density to 2.0×10^5 cells/mL and transfer 10 µL of transfected cells to each well of a white 384-well flat-bottom plate.
3. Dispense tracer at a concentration not exceeding 2× EC_{50} to each well, as determined in Sect. 3.5.2 (*see* **Note 16**).
4. Add increasing concentrations of peptidomimetic compounds to respective wells at a maximum DMSO concentration of 1% (*see* **Note 17**).
5. Add 5 µL Nano-Glo substrate diluted 1:200 in Opti-MEM to each well.
6. Measure BRET ratio using a microplate reader with a luminescence filter pair [450 nm BP filter and 610 nm LP filter].
7. Lyse cells by adding 25 nL of 20 mg/mL digitonin to each well, then incubate for 5 min.
8. Measure the BRET ratio again with the same settings as in Step 6 for permeabilized cells.

3.5.4 NanoBRET Data Analysis

The BRET ratio is calculated using the following equation:

$$BRET\ ratio\ [mBU] = \frac{Donor\ emission\ (450\ nm)}{Acceptor\ emission\ (610\ nm)} \times 1000$$

To determine the EC_{50} value of the tracer, the concentration is plotted against the baseline corrected BRET ratio and fitted with a non-linear model (Fig. 7A):

$$Y = \frac{X \times Y(Top)}{EC_{50} + X}$$

Y represents the BRET ratio, X denotes the tracer concentration, and Y(Top) corresponds to the BRET ratio at the highest compound concentration.

Displacement data is evaluated using normalized competition data with the following equation to determine IC_{50} values for each compound (Fig. 7B):

$$Y = \frac{100}{1 + \frac{X}{IC_{50}}}$$

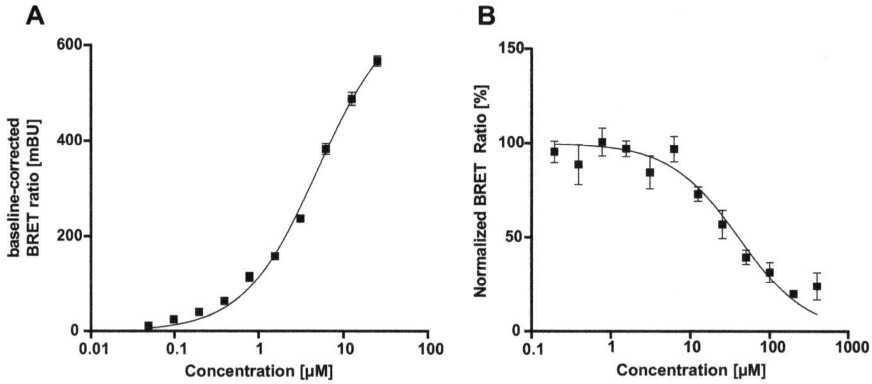

Fig. 7 (**A**) NanoBRET tracer titration of a BODIPY-conjugated peptidomimetic tracer in permeabilized HEK293T cells overexpressing full-length Nluc-TRIM7. (**B**) Competitive dose–response analysis showing displacement of the tracer molecule by the peptidomimetic compound in permeabilized cells. The respective non-linear regression fit is shown in black as overlaid curves

Here, X represents the concentration of the compound, and Y is the normalized BRET ratio.

4 Notes

1. DSF should not be confused with NanoDSF, as they utilize different strategies. NanoDSF monitors unfolding through tryptophan and tyrosine fluorescence, while DSF detects protein unfolding using selective dye fluorescence [24].
2. It is important to get a good fluorescent signal to determine a reliable Tm value for any E3 ligase. Varying the protein concentration (~1–20 μM) can help in finding a suitable fluorescence intensity for the screening of compounds with different binding affinities.
3. Try to determine the solubility of any peptidomimetic compound/peptide prior to the measurement to avoid precipitation effects.
4. DMSO concentration should be the same for all compounds and added to the "only protein" control.
5. Even though fluorescein isothiocyanate (FITC) and dyes with similar spectral properties are among the most widely used fluorophores, red dyes such as Cy5 have gained traction due to the improved performance of modern plate readers [25].
6. If the K_D of the peptide is already known, start with a tracer concentration lower than the K_D.

7. Concentrations of the compounds should cover at least 0.1× to 10× K_D of the compound-protein interaction to ensure enough curvature and the generation of valid plateaus for both maximal and minimal displacement.
8. Every solution and buffer, as well as the sensor chip, should be warmed up to RT prior to starting the assay to avoid baseline drifts and artifacts.
9. The buffer used to dilute samples needs to be exactly the same as the running buffer to avoid response deviations caused by differences in buffer composition.
10. By using a very low pH, you might risk a loss of protein integrity due to a sharp pH drop.
11. Protein levels should be selected according to reaching an expected maximum response (R_{max}) for interacting peptidomimetics of ~10–20 RU. Calculations can be done via:

$$Response(protein) = \frac{R_{max}(analyte) \times MW(protein)}{MW(analyte)}$$

12. For different DMSO concentrations, adjust both solutions to ±1% relative to the desired final DMSO concentration.
13. For functional kinetic evaluation, the association phase should exhibit visible curvature, while the dissociation phase should show at least 5% dissociation of the compound [26].
14. Opti-MEM medium should be free of PS and FBS to ensure optimal screening conditions, minimizing nonspecific interactions as well as high background fluorescence/luminescence signals.
15. Tracer concentrations should be chosen according to the expected K_D of its binding motif to cover 0.1× to 10× K_D. Moreover, a no-tracer control should be added for baseline correction.
16. For normalization purposes, include a no-tracer control and a sample containing the tracer but without the compound.
17. Select a broad concentration range of compound (0.1× to 10× the expected K_D) to ensure protein saturation and complete displacement of the tracer.

References

1. Sakamoto KM, Kim KB, Kumagai A et al (2001) Protacs: chimeric molecules that target proteins to the Skp1-Cullin-F box complex for ubiquitination and degradation. Proc Natl Acad Sci U S A 98:8554–8559. https://doi.org/10.1073/pnas.141230798
2. Zhong G, Chang X, Xie W et al (2024) Targeted protein degradation: advances in drug discovery and clinical practice. Signal Transduct Target Ther 9:308. https://doi.org/10.1038/s41392-024-02004-x

3. Békés M, Langley DR, Crews CM (2022) PROTAC targeted protein degraders: the past is prologue. Nat Rev Drug Discov 21:181–200. https://doi.org/10.1038/s41573-021-00371-6
4. Ishida T, Ciulli A (2021) E3 ligase ligands for PROTACs: how they were found and how to discover new ones. SLAS Discov 26:484–502. https://doi.org/10.1177/2472555220965528
5. Sharninghausen R, Hwang J, Dennison DD et al (2024) Identification of ERAD-dependent degrons for the endoplasmic reticulum lumen. eLife 12:RP89606. https://doi.org/10.7554/eLife.89606.3
6. Muñoz Sosa CJ, Lenz C, Hamann A et al (2024) A C-degron structure-based approach for the development of ligands targeting the E3 ligase TRIM7. ACS Chem Biol 19:1638–1647. https://doi.org/10.1021/acschembio.4c00301
7. Varshavsky A (2024) N-degron pathways. Proc Natl Acad Sci U S A 121:e2408697121. https://doi.org/10.1073/pnas.2408697121
8. Harris TJ, Trader DJ (2025) Exploration of degrons and their ability to mediate targeted protein degradation. RSC Med Chem 16:1067–1082. https://doi.org/10.1039/d4md00787e
9. Schwalm MP, Saxena K, Müller S et al. (2024) Luciferase- and HaloTag-based reporter assays to measure small-molecule-induced degradation pathway in living cells. Nat Protoc 19:2317–2357. https://doi.org/10.1038/s41596-024-00979-z
10. Pantoliano MW, Petrella EC, Kwasnoski JD et al. (2001) High-density miniaturized thermal shift assays as a general strategy for drug discovery. J Biomol Screen 6:429–440. https://doi.org/10.1177/108705710100600609
11. Lo M-C, Aulabaugh A, Jin G et al (2004) Evaluation of fluorescence-based thermal shift assays for hit identification in drug discovery. Anal Biochem 332:153–159. https://doi.org/10.1016/j.ab.2004.04.031
12. Checovich WJ, Bolger RE, Burke T (1995) Fluorescence polarization–a new tool for cell and molecular biology. Nature 375:254–256. https://doi.org/10.1038/375254a0
13. Nikolovska-Coleska Z, Wang R, Fang X et al (2004) Development and optimization of a binding assay for the XIAP BIR3 domain using fluorescence polarization. Anal Biochem 332:261–273. https://doi.org/10.1016/j.ab.2004.05.055
14. Nguyen HH, Park J, Kang S et al (2015) Surface plasmon resonance: a versatile technique for biosensor applications. Sensors (Basel, Switzerland) 15:10481–10510. https://doi.org/10.3390/s150510481
15. Tay L-L (2016) Surface plasmons. In: Luo MR (Ed.)Eds. Encyclopedia of color science and technology. Springer Science+Business Media, New York, 1186–1195
16. Tang Y, Zeng X, Liang J (2010) Surface plasmon resonance: an introduction to a surface spectroscopy technique. J Chem Educ 87:742–746. https://doi.org/10.1021/ed100186y
17. Roy MJ, Winkler S, Hughes SJ et al (2019) SPR-measured dissociation kinetics of PROTAC ternary complexes influence target degradation rate. ACS Chem Biol 14:361–368. https://doi.org/10.1021/acschembio.9b00092
18. Holdgate GA, Ward WHJ (2005) Measurements of binding thermodynamics in drug discovery. Drug Discov Today 10:1543–1550. https://doi.org/10.1016/S1359-6446(05)03610-X
19. Duff MR, Grubbs J, Howell EE (2011) Isothermal titration calorimetry for measuring macromolecule-ligand affinity. J Vis Exp 55:2796. https://doi.org/10.3791/2796
20. Johnson CM (2021) Isothermal titration calorimetry. Methods Mol Biol 2263:135–159. https://doi.org/10.1007/978-1-0716-1197-5_5
21. Frasca V (2016) Biophysical characterization of antibodies with isothermal titration calorimetry. J Appl Bioanal 2:90–102. https://doi.org/10.17145/jab.16.013
22. Machleidt T, Woodroofe CC, Schwinn MK et al (2015) NanoBRET–a novel BRET platform for the analysis of protein-protein interactions. ACS Chem Biol 10:1797–1804. https://doi.org/10.1021/acschembio.5b00143
23. Dale NC, Johnstone EKM, White CW et al (2019) NanoBRET: the bright future of proximity-based assays. Front Bioeng Biotechnol 7:56. https://doi.org/10.3389/fbioe.2019.00056
24. Wu T, Hornsby M, Zhu L et al (2023) Protocol for performing and optimizing differential scanning fluorimetry experiments. STAR Protoc 4:102688. https://doi.org/10.1016/j.xpro.2023.102688
25. (2025) Fluorescence polarization detection | BMG LABTECH. https://www.bmglabtech.com/en/fluorescence-polarization/. Accessed 04 Mar 2025
26. Katsamba PS, Navratilova I, Calderon-Cacia M et al (2006) Kinetic analysis of a high-affinity antibody/antigen interaction performed by multiple biacore users. Anal Biochem 352:208–221. https://doi.org/10.1016/j.ab.2006.01.034

Part III

Application of Degrons for Studying Biological Processes in Cells

Chapter 8

Multicolor Live-Cell mRNA Imaging Using RNA-Regulated Destabilization Domains

Tien G. Pham, Omoyemi Ajayi and Jiahui Wu

Abstract

mRNA molecules are critical for relaying genetic information from the genome into functional proteins for maintaining cellular function. Real-time imaging of mRNA in living cells reveals the spatiotemporal dynamics of RNA and offers insights into RNA-mediated regulation of gene expression and cellular function. Here, we describe a method of using RNA-regulated destabilization domains to simultaneously image multiple mRNAs in living cells. In this method, three mRNAs of interest are engineered to contain three distinct RNA imaging tags. Each tag specifically binds to and stabilizes its cognate RNA-regulated destabilization domain fused to fluorescent proteins, resulting in selective activation of fluorescence on the corresponding mRNA for imaging.

Keywords Live-cell RNA imaging, RNA-regulated destabilization domains, Fluorogenic proteins, RNA imaging tags, Fluorescence microscopy, Mammalian cells

1 Introduction

mRNA molecules are critical for relaying genetic information from the genome into functional proteins for maintaining cellular function. Importantly, the spatiotemporal distribution of mRNA within the cell is a major determinant of its function. For example, *CLB2* mRNA localized in the bud enhances protein translation in *Saccharomyces cerevisiae*, enabling cells to coordinate growth with cell cycle progression [1]. The *NET1* mRNA is prominently targeted to peripheral protrusions of migrating cells to promote cell adhesion and migration [2]. Additionally, *Nanos* mRNA localization to the posterior pole of the Drosophila embryo is crucial for control of gene activity and embryonic polarity [3]. Altogether, these highlight how spatial control of mRNAs can regulate fundamental cellular processes. Therefore, being able to track the spatial distribution of

The authors "Tien G. Pham and Omoyemi Ajayi" contributed equally.

Maria W. Górna et al. (eds.), *Degrons*, Methods in Molecular Biology, vol. 3069,
https://doi.org/10.1007/978-1-0716-5508-5_8,

mRNAs and its change over time are crucial for understanding how gene expression is regulated.

Many RNA imaging methods have been developed to enable tracking of the dynamics of RNA living cells [4–6]. For example, the RNA hairpin method based on MS2-MCP utilizes a naturally occurring bacteriophage-derived RNA hairpin, MS2, and its binding protein partner, MS2 coat protein (MCP), to tether fluorescent proteins to an mRNA of interest [7]. With 24 copies of MS2 appended to the 3' untranslated region (UTR) of mRNA of interest, *ASH1* mRNA localization and dynamics can be visualized through the aggregation of fluorescence signals at the RNA tag [7]. Alternatively, another method utilizes fluorogenic RNA aptamers to image RNA in living cells [8]. Fluorogenic RNA aptamers are engineered RNA sequences that can bind and turn on the fluorescence signals of otherwise non-fluorescent small-molecule dyes. When fused to an RNA of interest, these fluorogenic RNA aptamers can confer fluorescence signals to the RNA of interest for live-cell RNA imaging [8]. Despite rapid development on RNA imaging methods, simultaneously tracking more than two RNAs in living cells remains challenging.

Recently, we developed three orthogonal RNA-regulated destabilization domains for live-cell RNA imaging [9, 10]. In this method, we engineered three orthogonal RNA-regulated destabilization domains, and each one contains C-terminal Arg-Arg-Arg-Gly degron. When fused to fluorescent proteins, these fluorescent protein-destabilization domain fusions would be degraded in mammalian cells, thus conferring instability to the fluorescent proteins. However, when the cognate RNA binds to the destabilization domain, the binding of RNA can block the C-terminal degron from being recognized by the proteasomal degradation machinery, resulting in stabilization. By concatenating multiple repeats of each cognate RNA as a tag and fusing these tags to the 3'UTR of mRNAs, multiple mRNA species can be imaged (Fig. 1). Here, we describe a method of using these orthogonal RNA-regulated destabilization domains for three-color live-cell mRNA imaging.

2 Materials

2.1 Cell Culture

1. U2OS cells.
2. Cell culture medium: Phenol red-free DMEM supplemented with 10% fetal bovine serum, 100 U/mL of Penicillin, 100 μ g/mL of Streptomycin, 1× GlutaMAX™, and 1 mM sodium pyruvate.
3. T75 cell culture flask.
4. 1× PBS (Phosphate-Buffered Saline), pH 7.4: KH_2PO_4 (144.0 mg/L), NaCl (9000.0 mg/L), $Na_2HPO_4 \cdot 7H_2O$ (795.0 mg/L), premade by manufacturer.

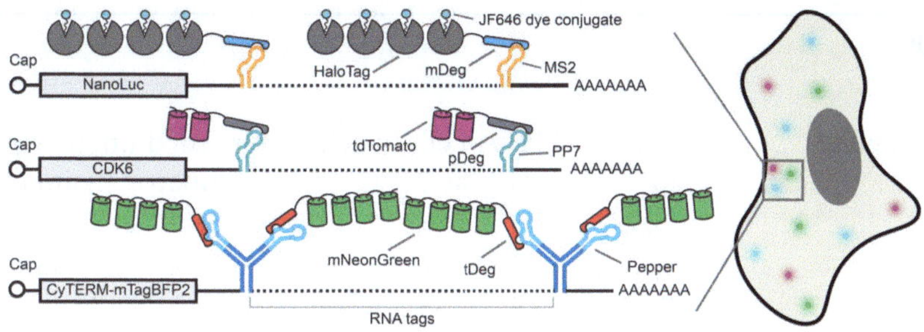

Fig. 1 Schematic of three-color mRNA imaging using orthogonal RNA-regulated destabilization domains. Three mRNAs of interest (NanoLuc, CDK6, and CyTERM-mTagBFP2) are engineered to contain three distinct RNA imaging tags [24XMS2, 24XPP7, and $(F30\text{-}2xPepper)_{10}$], respectively. Each tag specifically binds to and stabilizes its cognate RNA-regulated destabilization domain fused to fluorescent proteins. This results in selective activation of fluorescence on the corresponding mRNA for imaging

5. 1× TrypLE Express (no phenol red), premade by manufacturer.
6. 100× GlutaMAX™, premade by manufacturer.
7. 100 mM sodium pyruvate, premade by manufacturer.
8. 15 mL tube.
9. 50 mL tube.
10. Trypan Blue Solution, 0.4%, premade by manufacturer.
11. 35 mm imaging dish with glass bottom, precoated with poly-D-lysine by manufacturer.
12. Cell counting chamber slide

2.2 Transfection

1. Opti-MEM, premade by manufacturer (*see* **Note 1**).
2. DNA plasmids (*see* **Notes 2 and 3**):
   i. CMV-NanoLuc-24XMS2.
   ii. CMV-CDK6-24XPP7.
   iii. CMV-CyTERM-mTagBFP2-$(F30\text{-}2xPepper)_{10.}$
   iv. miniCMV-$(HaloTag)_4$-mDeg.
   v. miniCMV-tdTomato-pDeg.
   vi. miniCMV-$(mNeonGreen)_4$-tDeg.
3. FuGENE® HD transfection reagent, premade by manufacturer.

2.3 Imaging

1. Imaging medium: FluoroBrite™ (1×), premade by manufacturer.
2. Janelia Fluor® JF646 HaloTag ligand.
3. TYPE 37 immersion oil ($N = 1.515$), premade by manufacturer.

3 Methods

3.1 Cell Culture and Seeding

1. Culture U2OS cells in cell culture medium to 90% confluency in a T75 flask at 37°C in an incubator maintained with 5% CO_2.
2. Warm the cell culture medium, 1× PBS, and 1× TrypLE Express to 37°C in a water bath preheated to 37°C.
3. Aspirate the culture medium from the U2OS containing flask and gently rinse the cells once with 10 mL of prewarmed 1× PBS.
4. Aspirate the 1× PBS completely, then add 2 mL of 1× TrypLE Express to the flask.
5. Incubate 1× TrypLE Express with the U2OS cells for 5 min at 37°C in an incubator maintained with 5% CO_2 to allow cell detachment.
6. Once the cells detach, add 8 mL of prewarmed culture medium and gently pipette using a serological pipette to fully detach the cells.
7. Transfer the cell suspension to a 15 mL tube and centrifuge at 400 g for 4 min at 25°C to pellet the cells.
8. Aspirate the supernatant carefully, avoiding disturbance to the cell pellet.
9. Resuspend the cell pellet in 4 mL of prewarmed culture medium by gentle pipetting using a P1000 pipette then transfer the homogeneous suspension to a 50 mL tube.
10. Mix 15 μL of the homogeneous cell suspension with 15 μL of trypan blue solution, load 10 μL of this mixture onto a cell counting chamber slide, and count the cells.
11. Dilute the U2OS cell suspension in cell culture medium to a concentration of 1×10^6 cells/mL, based on the cell density calculated in Step 10.
12. Seed 2.7×10^5 cells in 2 mL of culture medium into a poly-D-lysine-coated 35 mm imaging dish, gently swirl to ensure even distribution and incubate overnight at 37°C with 5% CO_2 (*see* **Note 4**).

3.2 Transfection

1. Transfect U2OS cells 18–24 h after seeding onto the 35 mm imaging dish. For a single 35 mm imaging dish, combine the following reagents in a 1.7 mL tube: 80 μL of Opti-MEM, a total of 1.86 μg of DNA plasmids containing 0.31 μg of each plasmid listed in Sect. 2.2, and 6.4 μL of FuGENE® HD reagent (*see* **Note 5**).

2. Mix the mixture above thoroughly by gently pipetting, then incubate the mixture for 15 min at room temperature (20°C–25°C).
3. Add the transfection mixture to U2OS cells in a 35 mm imaging dish dropwise.
4. Swirl the dish gently to ensure homogeneous distribution of the transfection mixture in the cell culture medium.
5. Place the 35 mm imaging dish back in the incubator and culture the cells overnight at 37°C with 5% CO_2.
6. 18–24 h after transfection, aspirate the cell culture medium then rinse the cells with 2 mL of 1× PBS prewarmed to 37°C.
7. Aspirate the 1× PBS and add 2 mL of fresh cell culture medium prewarmed to 37°C and return the 35 mm imaging dish to the incubator. Culture the cells overnight at 37°C with 5% CO_2.

3.3 Imaging

1. Prewarm imaging medium in a water bath to 37°C.
2. Turn on the temperature control unit of the microscope imaging stage and allow the microscope imaging stage and environmental chamber to equilibrate to 37°C.
3. Thaw JF646 HaloTag ligand stored at −20°C to room temperature and resuspend in DMSO to generate a 200 μM stock solution.
4. Dilute the stock solution in the imaging medium and add it to U2OS cells to a final concentration of 200 nM. Incubate the cells with the JF646 HaloTag ligand at 37°C with 5% CO_2 for 30–60 min prior to imaging.
5. Aspirate the cell culture medium containing the JF646 HaloTag ligand from the 35 mm imaging dish, gently rinse the cells with 2 mL of prewarmed imaging medium. Aspirate the imaging medium, then add 2 mL of fresh, prewarmed imaging medium to the dish.
6. Apply enough immersion oil to a 60×/1.4-NA oil-immersion objective, then place the 35 mm imaging dish on the prewarmed microscope imaging stage.
7. Adjust the objective height to bring the adhered U2OS cells into focus using phase-contrast illumination.
8. Switch imaging mode from phase to fluorescence acquisition. Use a Cy5 filter cube (with excitation filter 620 ± 30 nm, dichroic mirror 660 nm (long pass), and emission filter 700 ± 37.5 nm), a dsRed filter cube (with excitation filter 545 ± 15 nm, dichroic mirror 570 nm (long pass), and emission filter 620 ± 30 nm), and a YFP filter cube (with excitation 500 ± 10 nm, dichroic mirror 515 nm (long pass), and emission filter 535 ± 15 nm), to identify transfected U2OS cells coexpressing the HaloTag

(with JF646 ligand), red fluorescent protein, tdTomato, and yellow fluorescent protein, mNeonGreen (*see* **Note 6**). Adjust exposure time and light intensity in the microscope software to acquire appropriate fluorescence images (*see* **Note 7**).

4 Notes

1. OptiMEM can be substituted with double-distilled water for FuGENE® HD-mediated transfection.
2. DNA plasmids containing long repetitive sequences are prone to recombination and deletion during plasmid propagation. To minimize this, we suggest propagating these plasmids using XL10-Gold *E.coli* cells or the NEB stable *E. coli* cells.
3. Different mRNA reporters have different cellular localizations: NanoLuc-24XMS2 and CDK6-24XPP7 mRNAs localize in the cytosol; CyTERM-mTagBFP2-(F30-2xPepper)$_{10}$ mRNA localizes on the endoplasmic reticulum (ER) membrane.
4. Typically, for tracking single mRNA molecules, 2.7×10^5 U2OS cells were cultured in a 35 mm dish in 2 mL of culture medium. For different experiments, depending on the cell types, an optimized number of cells may be used.
5. Careful optimization of transfection parameters, including the amount of total DNA plasmids and the DNA plasmid-to-transfection reagent ratio are crucial for efficient live cell imaging. Generally, we use a total of 1.86 μg of DNA plasmids containing 0.31 μg of each plasmid listed in Sect. 2.2.
6. To avoid phototoxicity and fluorophore photobleaching during live-cell imaging, we suggest using low excitation power of the light source when moving across the sample to identify transfected cells.
7. The excitation power of the light source and the camera acquisition time should be adjusted to balance signal-to-noise ratio with photobleaching during mRNA imaging. If the fluorescence signal from the mRNA is too low, consider removing all neutral-density filters from the excitation light source to maximize the excitation intensity. In addition, because the fluorescence signal from single mRNA molecules is inherently weak, we recommend using a sensitive camera, such as an EMCCD or sCMOS camera, for image acquisition or movie recording.

Acknowledgments

This work was supported by the UMass startup funds and NIH R35GM160207.

References

1. Maekiniemi A, Savakis P, van Rossum K, Snoep JL, Seiler M, van Niekerk DD, Zarnack K, Singer RH, Tutucci E (2025) Cyclin CLB2 mRNA localization and protein synthesis link cell cycle progression to bud growth. Nat Commun 16:11654. https://doi.org/10.1038/s41467-025-66623-w
2. Gasparski AN, Moissoglu K, Pallikkuth S, Meydan S, Guydosh NR, Mili S (2023) mRNA location and translation rate determine protein targeting to dual destinations. Mol Cell 83:2726–2738.e9. https://doi.org/10.1016/j.molcel.2023.06.036
3. Gavis ER, Lehmann R (1992) Localization of nanos RNA controls embryonic polarity. Cell 71:301–313. https://doi.org/10.1016/0092-8674(92)90358-J
4. Urbanek MO, Galka-Marciniak P, Olejniczak M, Krzyzosiak WJ (2014) RNA imaging in living cells: methods and applications. RNA Biol 11:1083–1095. https://doi.org/10.4161/rna.35506
5. Pham TG, Wu J (2024) Recent advances in methods for live-cell RNA imaging. Nanoscale 16:5537–5545. https://doi.org/10.1039/D4NR00129J
6. Mi L, Khajouei S, You M (2025) Multiplexed RNA imaging and in situ profiling in living cells. Chem Sci 16:21152–21173. https://doi.org/10.1039/D5SC06220A
7. Bertrand E, Chartrand P, Schaefer M, Shenoy SM, Singer RH, Long RM (1998) Localization of ASH1 mRNA particles in living yeast. Mol Cell 2:437–445. https://doi.org/10.1016/S1097-2765(00)80143-4
8. Paige JS, Wu KY, Jaffrey SR (2011) RNA mimics of green fluorescent protein. Science 333:642–646. https://doi.org/10.1126/science.1207339
9. Wu J, Zaccara S, Khuperkar D, Kim H, Tanenbaum ME, Jaffrey SR (2019) Live imaging of mRNA using RNA-stabilized fluorogenic proteins. Nat Methods 16:862–865. https://doi.org/10.1038/s41592-019-0531-7
10. Pham TG, Ajayi O, He J, Sagarbarria I, Hardy JA, Wu J (2025) Orthogonal RNA-regulated destabilization domains for three-color RNA imaging with minimal RNA perturbation. Nat Methods 23:165–174. https://doi.org/10.1038/s41592-025-02905-x

Chapter 9

Dynamic Control of Cell Signaling with Optogenetic Modulation of Protein Abundance in Living Mammalian Cells

Yu-En Huang, Payel Mondal, Huaxun Fan and Kai Zhang

Abstract

Precise modulation of protein levels is essential for investigating the dynamics of signaling pathways. Degradation-based approaches allow rapid reduction of protein abundance, and their combination with optogenetic regulation enables controlled and reversible manipulation. In this chapter, we describe an optogenetic strategy that stabilizes a protein of interest by inhibiting its degradation. As a proof of concept, we demonstrate light-dependent stabilization of MKP3 and constitutively active MEK and examine the resulting changes in ERK signaling. We also provide a step-by-step protocol for measuring protein stabilization kinetics. This method integrates a potent degradation tag with optogenetic control to achieve tunable and reversible regulation of protein levels in living cells.

Keywords Optogenetics, GLIMPSe, MKP3, CA MEK, PC12 cells, Neurite outgrowth

1 Introduction

Cellular signaling is governed by an intricate network of interconnected pathways. Rather than functioning in a simple binary on/off fashion, these pathways often exhibit highly dynamic behaviors. An essential driving force behind this dynamic nature is the spatial and temporal control of protein activities. To manipulate these protein activities and signaling pathways, genetic approaches, such as CRISPR knockout and RNA interference (RNAi), have been developed to control the levels of protein targets [1, 2]. These methods are effective at suppressing the synthesis of protein targets. However, they have no effect on existing proteins, leading to their ineffectiveness against proteins that have long half-lives [3, 4]. They are also susceptible to off-target effects, and cells often develop compensatory pathways when a gene is disrupted or removed [5–9]. Another obstacle for genetic approaches is that they lack the flexibility to capture signaling dynamics. To better study the dynamic nature of cellular events, a rapid and reversible post-translational method is needed.

Maria W. Górna et al. (eds.), *Degrons*, Methods in Molecular Biology, vol. 3069,
https://doi.org/10.1007/978-1-0716-5508-5_9,

The protein degradation approach offers an alternative strategy to overcome challenges from genetic approaches. By acting directly on existing proteins, it can deplete the target's function without a long lag associated with genetic knock-out or knock-down procedures [3, 10–14]. At the same time, because degradation removes proteins post-translationally without altering their ongoing synthesis, it allows rapid restoration once the degradation cue is removed [13]. Induced degradation of a protein can be achieved by attaching a degron, such as the ubiquitin-independent Dnd1 degron, which potently reduces target protein abundance [14, 15]. However, because degradation is inherently continuous, an external layer of control is required to make this process dynamic and tunable. Optogenetic regulation, which uses light as a precise, non-invasive stimulus, provides a powerful means to impose spatiotemporal control [16, 17]. Integrating an optogenetic module with a degradation-based system would therefore enable highly precise regulation of protein activity and the signaling pathways in which they function.

In this chapter, we introduce a protein degradation-based platform termed the generalizable light-modulated protein stabilization system (GLIMPSe) [14]. This method achieves dynamic control of protein activity by combining protein degradation with optogenetics (Fig. 1). The GLIMPSe system consists of the protein of interest (POI) fused to a light-responsive eLOV domain and three copies of the Dnd1 degron. The eLOV domain cages the TEV protease recognition site (tevS). Under normal, unstimulated circumstances, the degrons mediate degradation of POI, leading to a constitutively suppressed activity. When blue light illumination begins, the caged tevS is exposed for cleavage. Concurrently, the TEV protease, fused to a light-inducible nuclear export system (LEXY), is only exported from the nucleus into the cytosol when exposed to blue light. The

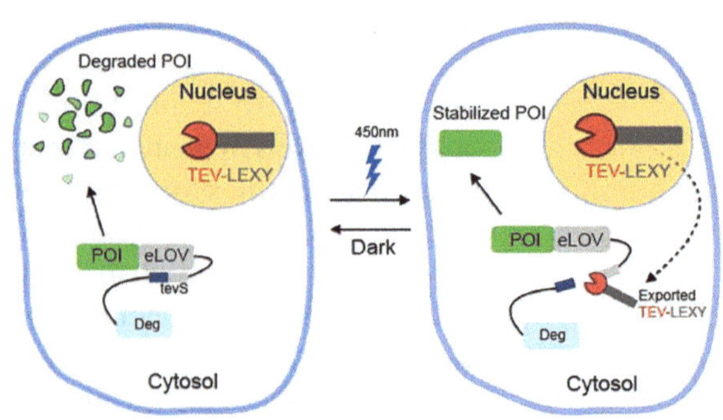

Fig. 1 Schematic outline of GLIMPSe. The GLIMPSe platform consists of the protein of interest (POI) fused to an eLOV domain caging tevS and Dnd1 degron, as well as TEV-LEXY that is sequestered to the nucleus. When there is no light stimulation, the degron sequence mediates degradation of the POI. Upon exposure to blue light, eLOV uncages tevS, and TEV-LEXY is exported into the cytosol. The degron sequence is then cleaved from the POI, achieving POI stabilization

protease would then cut the fusion protein at tevS and remove the degron sequence, stopping the degradation of POI. The two light-induced events combined restore and stabilize the POI level. Here, we demonstrate bidirectional manipulation of the extracellular signal-regulated kinase (ERK) signaling pathway by either stabilizing mitogen-activated protein kinase phosphatase 3 (MKP3) or constitutively active dual specific mitogen-activated protein kinase kinase (CA MEK) and their effect on neurite outgrowth of PC12 cells.

2 Materials

2.1 Equipment

1. BSL2 tissue culture biosafety cabinet.
2. Tissue culture incubator at 37°C with 5% CO_2.
3. A computer-controlled, programmable home-built light box designed to supply both pulsed and continuous blue light. This is constructed by assembling a 6-by-4 blue LED array with 24 blue LEDs (B4304H96, Linrose Electronics) on a breadboard. The breadboard is embedded in an aluminum box. A light diffuser film is placed on top of the breadboard to homogenize the light intensity.

2.2 Cell Transfection

1. PC12 cells.
2. Growth media (F12K media, supplemented with 15% horse serum (HS) and 2.5% fetal bovine serum (FBS)).
3. Starvation media (F12K media, supplemented with 0.15% HS and 0.025% FBS).
4. Transfection reagent such as Turbofect.
5. 12-well plate.
6. CA MEK-EGFP plasmid. Mutations were introduced via site-directed mutagenesis. CA-MEK is a constitutively active MEK that contains S218D and S222D mutations [14] (*see* **Note 1**).
7. EGFP-p2A-MKP3-eLOV-tevS-deg plasmid (*see* **Notes 1 and 2**).
8. EGFP-p2A-CA MEK-eLOV-tevS-deg plasmid (*see* **Notes 1 and 2**).
9. NLS-mCherry-TEV-LEXY plasmid (*see* **Note 1**).

2.3 Western Blot

1. PVDF membrane.
2. RIPA lysis buffer (10× RIPA lysis buffer stock + water + protease and phosphatase inhibitor cocktail).
3. Protease and phosphatase inhibitor cocktail.

4. SDS-PAGE running buffer (25 mM Tris, 192 mM glycine, 0.1% SDS, pH 8.3).
5. ECL substrate.
6. LDS loading buffer.
7. β-Mercaptoethanol.
8. Precast SDS-PAGE gel.
9. Transfer buffer (25 mM Tris, 192 mM glycine, 20% methanol, pH 8.3).
10. Phospho-p44/42 MAPK antibody.
11. P44/42 MAPK antibody.
12. HA-tag antibody.
13. HRP-linked anti-rabbit IgG secondary antibody.

3 Methods

3.1 Inhibit pERK Through Optogenetic Stabilization of MKP3

1. Seed PC12 cells in F12K growth media on two 12-well plates (*see* **Note 3**).
2. Let the cells grow to 40%–50% confluency.
3. Perform transfection with Turbofect transfection reagent following the vendor's instructions. For optogenetic stabilization of MKP3, transfect 100 ng CA MEK-EGFP, 200 ng EGFP-p2A-MKP3-eLOV-tevS-deg, and 700 ng of NLS-mCherry-TEV-LEXY plasmids into the cell (*see* **Notes 4–6**).
4. After 3 h of transfection, replace the media in the well with F12K growth media.
5. Place one plate on the light box and move them into the incubator. Begin blue light stimulation at an intensity of 0.2–0.5 mW/cm^2. Keep the other plate in the dark as a negative control (*see* Notes **7 and 8**).
6. Stimulate the PC12 cells for 48–72 h before imaging neurite outgrowth.
7. Image the cells to determine the differentiation ratio based on the equation below (*see* Notes **9–11**).

$$\textit{Neurite outgrowth ratio} = \frac{\#\textit{of mCherry-positive cells showing neurite outgrowth}}{\#\textit{of EGFP-positive cells}}$$

Representative fluorescence images of PC12 cells transfected with GLIMPSe-MKP3 are shown in Fig. 2. Significant neu-

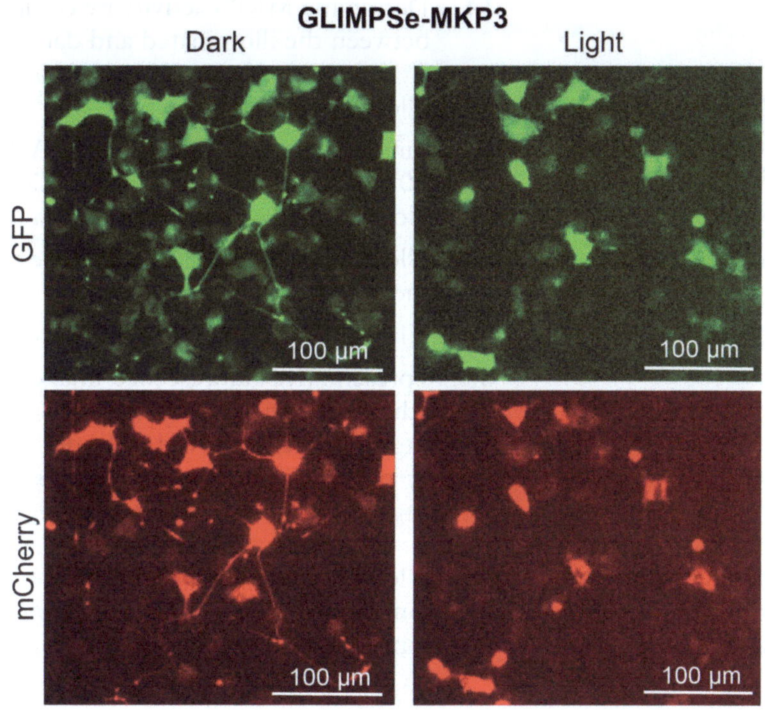

Fig. 2 GLIMPSe-mediated stabilization of MKP3 inhibits differentiation of PC12 cells. PC12 cells were transfected with CA-MEK-GFP, GFP-2A-MKP3-eLOV-tevS-degron, and NLS-mCherry-TEV-LEXY. When cells were left in the dark, robust signs of neurite outgrowth, a feature of differentiation, can be observed. When cells are stimulated by blue light, this feature is gone

rite outgrowth was observed in the dark (MKP3 degraded, Fig. 2 left panels), whereas blue light stimulation blocked neurite outgrowth (MKP3 restored, Fig. 2, right panels). Follow Steps 9–15 if performing a Western blot.

8. Harvest and lyse the cell with 1× RIPA buffer supplemented with protease/phosphatase inhibitor cocktail.
9. Centrifuge the lysate at 15,000 × *g* for 10 min to clear away cell debris.
10. Mix the lysate with NuPAGE LDS Sample Buffer and β -Mercaptoethanol.
11. Run SDS-PAGE on a precast gel, followed by overnight transfer to a PVDF membrane.
12. Perform primary blotting with an antibody for phospho-ERK and pan ERK.
13. Visualized secondary blotting with the corresponding secondary antibody.

14. Determine MKP3 activity by comparing the pERK/panERK between the illuminated and dark plates (*see* **Note 12**).

3.2 Promote pERK Through Optogenetic Stabilization of CA-MEK

1. Follow Steps 1 and 2 of Sect. 3.1.
2. Transfect 50 ng of EGFP-p2A-CA MEK-eLOV-tevS-deg and 950 ng of NLS-mCherry-TEV-LEXY plasmids using Turbofect following the vendor's protocol (*see* **Notes 4–6**).
3. Follow Steps 4–7 of Sect. 3.1. The total blue light stimulation time should be around 45 h.
4. Follow up with imaging or a Western blot experiment.
5. Representative fluorescence images of PC12 cells transfected with GLIMPSe-CA-MEK are shown in Fig. 3. Minimal neurite outgrowth was observed in the dark (CA-MEK degraded, Fig. 3 left panels), whereas blue light stimulation induced neurite outgrowth (CA-MEK restored, Fig. 3, right panels).

3.3 Kinetic Analysis of GLIMPSe-Mediated MKP3 Stabilization

1. Follow Steps 1-2 of Sect. 3.1. Plate one well of cells for each illumination time point. For example, to measure seven time points (0, 0.5, 1, 2, 4, 6, and 24 h), prepare seven wells per

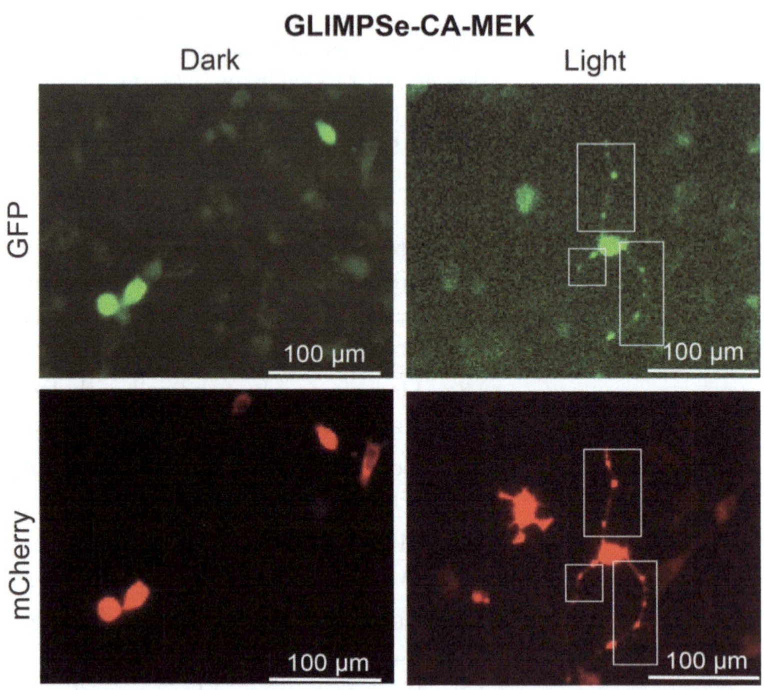

Fig. 3 GLIMPSe-mediated stabilization of CA MEK promotes differentiation of PC12 cells. PC12 cells were transfected with GFP-2A-CA MEK-eLOV-tevS-degron and NLS-mCherry-TEV-LEXY. When cells were left in the dark, no signs of neurite outgrowth were observed. When cells are stimulated by blue light, neurite outgrowth is observed, as shown in the white boxes

biological replicate, with each well harvested and analyzed at its designated time point.

2. Transfect 100 ng of EGFP-p2A-HA-MKP3-eLOV-tevS-deg and 50 ng of NLS-mCherry-TEV-LEXY plasmids using Turbofect following the vendor's protocol (*see* **Notes 4–6**).
3. After 3 h of transfection, let the cells recover in F12K growth media overnight. Do not perform any stimulation yet.
4. The next day, expose the cells to blue light at an intensity of 0.5 mW/cm^2 for timespans ranging from 0 to 24 h. Use 0 h as the unstimulated negative control (*see* **Notes 7 and 8**).
5. Harvest cell lysates at the corresponding time points.
6. Follow up with Western blot using HA tag and GAPDH primary antibodies. On the HA blot, there should be two bands. One is the full-length construct at 72 kDa, and the other, with the degron sequences cleaved off, is at 62 kDa.
7. Determine the stabilization efficiency at a given time point t using the following equation

$$Stabilization\ efficiency\ (t) = \frac{I_{62\ kDa}(t)}{I_{62\ kDa}(t) + I_{72\ kDa}(t)}$$

Where I stands for the HA band intensity at their corresponding molecular weight.

Fit the curve to obtain the kinetics of protein stabilization.

4 Notes

1. The plasmids listed here are in the process of being deposited to Addgene and are available upon request.
2. The full sequence of the caged degron (eLOV-tevS-deg) can be found in the supplementary information of Mondal et al. [14].
3. PC12 cells are easy to detach from surface. Use mild pipetting when switching medium or performing transfection.
4. Plasmids quality affects transfection efficiency. Using high-quality plasmids purified by Maxiprep helps achieve high transfection efficiency in PC12 cells.
5. Transfection efficiency differs for different cell lines. If chemical transfection results in poor transfection efficiency, consider lentiviral infection instead.
6. For each experiment, titrate the total amount and ratio of plasmids used to optimize the dynamic range (light-to-dark contrast) of neurite outgrowth ratio.

7. LEDs in the home-built lightbox can be connected with thin copper wires to facilitate illumination within the incubator.
8. If continuous light illumination results in strong phototoxicity, consider lowering the excitation power or use intermittent light pattern. In our hands, light power of 0.2-0.5 mW/ cm^2 is sufficient to activate the GLIMPSe system.
9. Tune cell plating density so that cells will not overgrow by the time of imaging.
10. If performing live-cell imaging, pre-equalize the environmental chamber at 37 °C with 5% CO_2 for at least 30 min before imaging.
11. Kinetics of the nuclear to cytosolic transport of TEV-LEXY can be measured by fluorescence imaging using mCherry channels.
12. If performing Western blot, normalize pERK activity with pan-ERK.

Acknowledgments

This work was supported by the National Institute of Child Health and Human Development Training Grant T32 HD108075 (Y.H.), the National Institute of Mental Health (R01MH124827), NSF (award No. 2121003), and an NSF Science and Technology Center for Quantitative Cell Biology grant (award No. 2243257) (K.Z.).

References

1. Doudna JA, Charpentier E (2014) The new frontier of genome engineering with CRISPR-Cas9. Science 346:1258096. https://doi.org/10.1126/science.1258096
2. Elbashir SM, Harborth J, Lendeckel W et al (2001) Duplexes of 21-nucleotide RNAs mediate RNA interference in cultured mammalian cells. Nature 411:494–498. https://doi.org/10.1038/35078107
3. Sun W, Zhang W, Zhang C et al (2017) Light-induced protein degradation in human-derived cells. Biochem Biophys Res Commun 487:241–246. https://doi.org/10.1016/j.bbrc.2017.04.041
4. Smoak EM, Stein P, Schultz RM et al (2016) Long-term retention of CENP-A nucleosomes in mammalian oocytes underpins transgenerational inheritance of centromere identity. Curr Biol 26:1110–1116. https://doi.org/10.1016/j.cub.2016.02.061
5. Guo C, Ma X, Gao F, Guo Y (2023) Off-target effects in CRISPR/Cas9 gene editing. Front Bioeng Biotechnol 11:1143157. https://doi.org/10.3389/fbioe.2023.1143157
6. Neumeier J, Meister G (2021) siRNA specificity: RNAi mechanisms and strategies to reduce off-target effects. Front Plant Sci 11:526455. https://doi.org/10.3389/fpls.2020.526455
7. El-Brolosy MA, Stainier DYR (2017) Genetic compensation: a phenomenon in search of mechanisms. PLoS Genet 13:e1006780. https://doi.org/10.1371/journal.pgen.1006780
8. Rossi A, Kontarakis Z, Gerri C et al (2015) Genetic compensation induced by deleterious mutations but not gene knockdowns. Nature 524:230–233. https://doi.org/10.1038/nature14580
9. Serobyan V, Kontarakis Z, El-Brolosy MA et al. (2020) Transcriptional adaptation in Caenorhabditis elegans. eLife 9:e50014. https://doi.org/10.7554/eLife.50014

10. Zhao L, Zhao J, Zhong K et al (2022) Targeted protein degradation: mechanisms, strategies and application. Sig Transduct Target Ther 7:113. https://doi.org/10.1038/s41392-022-00966-4
11. Sakamoto KM, Kim KB, Kumagai A et al (2001) Protacs: chimeric molecules that target proteins to the Skp1–Cullin–F box complex for ubiquitination and degradation. Proc Natl Acad Sci U S A 98:8554–8559. https://doi.org/10.1073/pnas.141230798
12. Clift D, McEwan WA, Labzin LI et al (2017) A method for the acute and rapid degradation of endogenous proteins. Cell 171:1692–1706.e18. https://doi.org/10.1016/j.cell.2017.10.033
13. Nabet B, Roberts JM, Buckley DL et al (2018) The dTAG system for immediate and target-specific protein degradation. Nat Chem Biol 14:431–441. https://doi.org/10.1038/s41589-018-0021-8
14. Mondal P, Krishnamurthy VV, Sharum SR et al (2019) Repurposing protein degradation for optogenetic modulation of protein activities. ACS Synth Biol 8:2585–2592. https://doi.org/10.1021/acssynbio.9b00285
15. Hwang H, Jin Z, Krishnamurthy VV, et al (2019) Novel functions of the ubiquitin-independent proteasome system in regulating Xenopus germline development. Development 146(8), dev172700. https://doi.org/10.1242/dev.172700
16. Dagliyan O, Hahn KM (2019) Controlling protein conformation with light. Curr Opin Struct Biol 57:17–22. https://doi.org/10.1016/j.sbi.2019.01.012
17. Zhang K, Cui B (2015) Optogenetic control of intracellular signaling pathways. Trends Biotechnol 33:92–100. https://doi.org/10.1016/j.tibtech.2014.11.007

Chapter 10

Reporter Gene Assay for Monitoring BCL6 Reactivation

Catherine L. Ingram, Radosław P. Nowak and Thomas M. Geiger

Abstract

Transcription factors are an exciting class of drug targets. Recent developments in small molecule-induced proximity allow for small molecule control of their activity by recruitment of epigenetic proteins. In this chapter, we describe the development of a transcriptional activity assay for the transcription factor BCL6 with a small molecule activator, TCIP1, in HEK293T cells [1]. The assay is based on the Bio-Glo Luciferase assay system and uses transient transfection of a BCL6-expressing plasmid, as well as a reporter plasmid where a BCL6 DNA-binding sequence controls expression of a luciferase protein. The method described here is specific to BCL6 but can easily be optimized to monitor the activity of other transcription factors by altering the DNA-binding sequence of the reporter.

Keywords BCL6, TCIP, Transcription activation, Induced proximity, Reporter gene assay

1 Introduction

Reporter gene assays provide a powerful tool to monitor the regulatory activity of transcription factors (TF) and activation of signal pathways [2, 3]. In these assays, the TF's activity is linked to a detectable signal by placing regulatory DNA elements (response elements) upstream of a reporter gene, such as a luciferase. When the transcription factor is active, the reporter gene is expressed, and the resulting signal can easily be quantified.

In drug discovery, reporter gene assays are commonly used to measure interference of protein ligands with defined signaling pathways in a rapid, sensitive, and scalable manner. Recently, a new class of hetero-bifunctional molecules has been developed that recruit transcriptional activators to repressors of promoters of cell death genes [1, 4–6]. These compounds, termed TCIPs (transcriptional and epigenetic chemical inducers of proximity), overwrite the repressive effect of oncogenic proteins, such as BCL6 (B-cell lymphoma 6 protein), ultimately leading to reactivation of BCL6-suppressed apoptotic genes and cell death through recruitment of transcriptional activators, such as BRD4 [1].

Maria W. Górna et al. (eds.), *Degrons*, Methods in Molecular Biology, vol. 3069,
https://doi.org/10.1007/978-1-0716-5508-5_10,

Here, we describe a detailed protocol for the implementation of a reporter gene assay for BCL6 utilizing transient transfection and HEK293T cells. Towards this aim, we generated a reporter plasmid containing 9 BCL6 DNA-binding elements (5′-TTCCTAGAA-3′) positioned upstream of a minimal promoter and the firefly luciferase gene (Fig. 1). The dTAG system is a combination of an FKBP12 protein with the F36V mutation and a corresponding small molecule degrader, such as dTAG13 (recruiting E3 ligase adapter CRBN) [7] or dTAG-V1 (recruiting E3 ligase adapter VHL) [8], which, when fused to a protein of interest, results in its selective degradation. Recently, the FKBP12^{F36V} system has been repurposed for recruitment of proteins other than E3 ligases that allow for transcriptional activation or protein relocalization [9]. Here, we leverage the BCL6-FKBP12^{F36V} system to activate transcription of BCL6 with TCIP1 [1], which recruits BRD4 directly to the BTB domain of BCL6, or with NICE-1 [9], which recruits BRD4 to the FKBP12^{F36V} part of the BCL6-FKBP12^{F36V} fusion. Upon transient co-transfection of the reporter plasmid together with a BCL6-FKBP12^{F36V} expression plasmid into HEK293T cells, BCL6 binds to its cognate DNA-binding elements upstream of the reporter gene, enabling the monitoring of transcriptional activation mediated by TCIP1 and NICE-1. These recruitments both result in activation of the reporter construct and increased firefly luciferase expression.

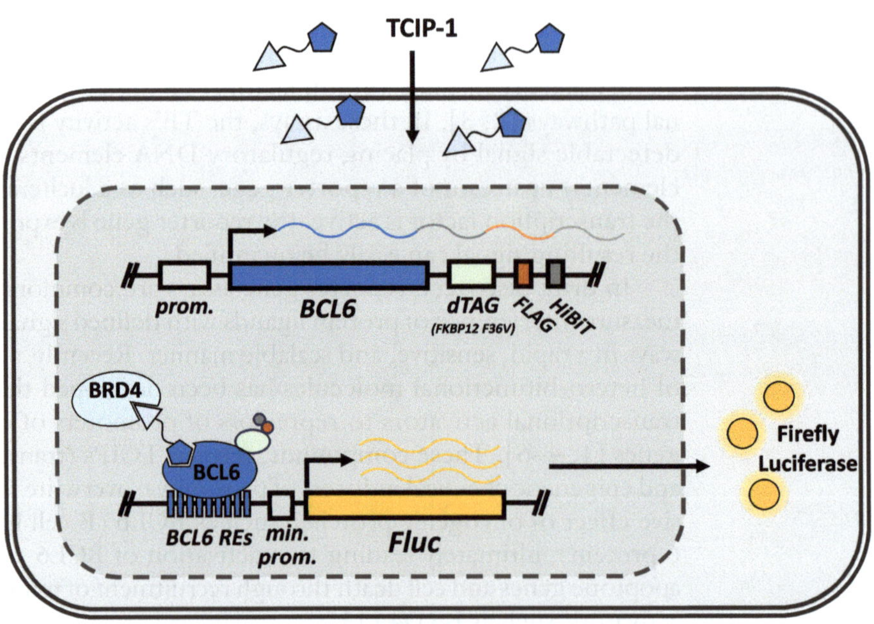

Fig. 1 Schematic visualization of the BCL6 reporter gene assay in HEK293T cells. The BCL6 reporter plasmid and a BCL6-dTAG expression construct are introduced by transient transfection. Upon expression of the BCL6-dTAG, small molecule TCIPs recruit endogenous BRD4 to DNA-bound BCL6, which increases firefly luciferase (Fluc) expression

Furthermore, a BCL6ΔZF- FKBP12^{F36V} construct lacking the zinc finger DNA-binding domain was used to demonstrate that the observed activation depends on BCL6 binding to the reporter vector and is therefore a consequence of BCL6- and TCIP-mediated proximity between BRD4 and the reporter construct. This system, relying on co-transfection of a reporter plasmid and a plasmid expressing a transcription factor of interest, can be easily adapted to monitor the activity of other transcription factors for which the DNA response elements are known.

2 Materials

2.1 Equipment

1. Cell culture equipment and reagents—include a biosafety cabinet and a heated bead/water bath.
2. Automated cell counter or hemocytometer.
3. Electronic multichannel pipette with adjustable tip spacing, such as the E1 Clip-Tip™ P125 multichannel pipette.
4. Automatic compound dispensing system (*see* **Note 1**).
5. Multi-well plate reader equipped with a luminescence module, such as a PHERAstar FSX (*see* **Note 2**).

2.2 Reagents

1. HEK293T/17 cell line.
2. DMEM, high glucose supplemented with 10% fetal bovine serum and 1% Penicillin-Streptomycin; *referred to here as culture medium*.
3. Opti-MEM I reduced serum medium, no phenol red.
4. Cell culture phosphate-buffered saline (1×).
5. Trypsin-EDTA (0.05%), phenol red.
6. Trypan Blue solution (0.4%).
7. Tissue culture (TC) dish 100 mm, standard, TC-treated.
8. White/opaque 384-well plates, sterile.
9. Transfection reagent, such as Lipofectamine™ 2000 or equivalent.
10. Firefly luciferase detection reagent, such as the Bio-Glo Luciferase Assay System (Promega Corporation catalog number G7940).
11. 9xRE (BCL6) reporter in pGL4.38 plasmid backbone reconstituted in DI water (see plasmid map in Fig. 2).
12. BCL6-dTAG-Flag-HiBiT in the Cilantro2 plasmid backbone lacking GFP and mCherry was reconstituted in DI water (see plasmid map in Fig. 3) (*see* **Note 3**).

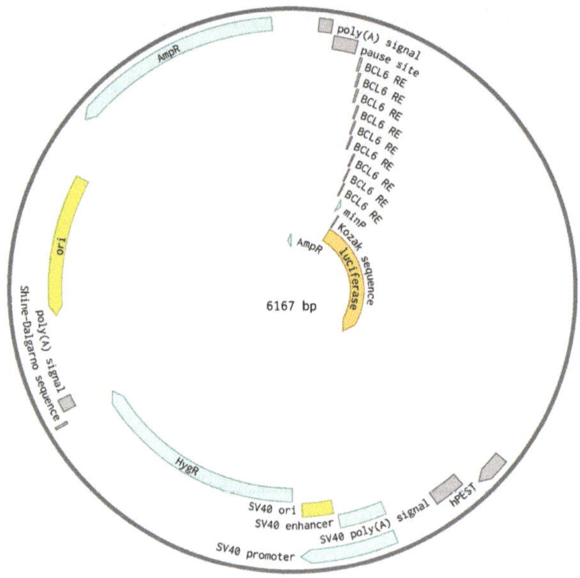

Fig. 2 Map of the BCL6 reporter gene plasmid. 9 BCL6 response elements (RE) are placed upstream of the firefly luciferase gene in the pGL4.38 backbone. The BCL6 response element (RE) corresponds to the sequence: TTCCTAGAA

13. Positive control compound TCIP1 [1]. Best to use fresh each time with *no freeze-thaws*; suggested to prepare a 10 mM stock in DMSO and store small aliquots at −80°C.
14. Compound NICE-01 [9]. Best to use fresh each time with *no freeze-thaws*; suggested to prepare a 10 mM stock in DMSO and store small aliquots at −80°C.

3 Methods

3.1 Transfection of Luciferase Reporter and BCL6-dTAG Expressing Plasmids

1. Culture HEK293T/17 in 100 mm TC plates under normal conditions as recommended by ATCC.
2. After collecting the confluent cells (*see* **Note 4**) in cell culture medium, prepare a 1:1 mixture of trypan blue to cell suspension. Use this suspension to count the cells and confirm viability of at least 90%.
3. Seed three 100 mm TC plates with 2.5×10^6 viable cells in 10 mL of cell culture medium per plate.
4. Place the plates in a mammalian cell culture incubator (37°C, 5% CO_2) overnight.
5. The following day, prepare the following solutions in 1 mL Opti-MEM in sterile 2 mL canonical tubes:

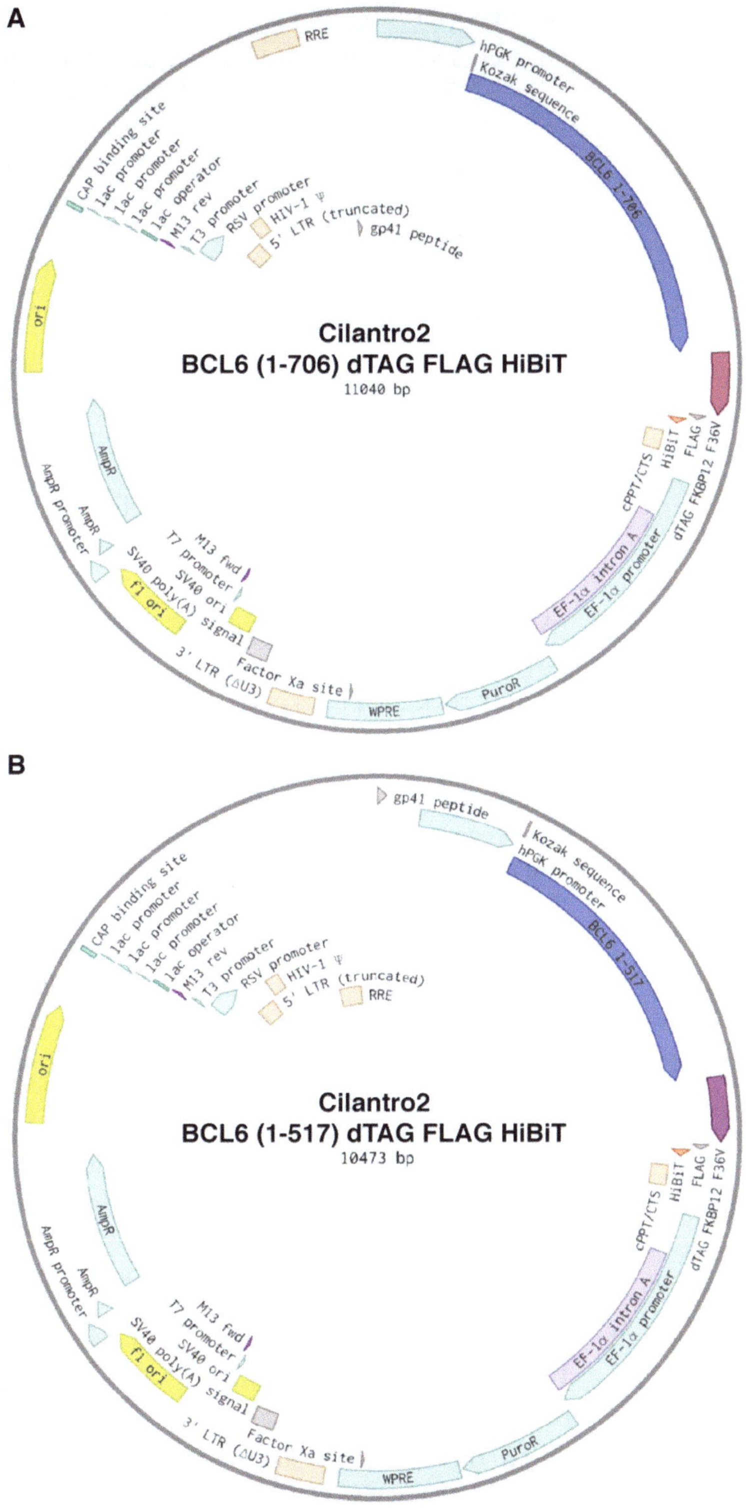

Fig. 3 Map of the plasmid overexpressing BCL6-dTAG. C-terminal BCL6 dTAG-Flag-HiBiT fusion constructs in the Cilantro2 backbone. (**A**) BCL6 (1-706) corresponds to full-length BCL6, while (**B**) BCL6ΔZF (1-517) lacks the C2H2 zinc finger domains

   a. Reporter: 5 μg BCL6 reporter vector.
   b. Reporter + BCL6-dTAG: 5 μg BCL6 reporter vector + 2.5 μg BCL6 (1-706)-dTAG-Flag-HiBiT expression vector.
   c. Reporter + BCL6ΔZF-dTAG: 5 μg BCL6 reporter vector + 2.5 μg BCL6ΔZF (1-517)-dTAG-Flag-HiBiT expression vector.

6. Prepare 50 μL Lipofectamine2000 in 3 mL Opti-MEM reduced serum medium in a separate tube and incubate for 5 min at RT.
7. Combine the 1 mL of the plasmid solutions with 1 mL of lipofectamine solution, respectively, and gently mix by pipetting up and down five times.
8. Incubate at room temperature for 20 min.
9. Aspirate the medium from each plate and replace it with 8 mL of fresh cell culture media.
10. Add 2 mL of the transfection mix dropwise to each plate of cells, respectively.
11. Gently rock the plates back and forth and place them back in the incubator for 24 h.

3.2 Plating and Treatment

1. Collect the transfected cells in culture medium (*see* **Note 4**) and prepare a 1:1 mixture of trypan blue to cell suspension. Use this suspension to count the cells and confirm viability of at least 90%.
2. Dilute to a final cell concentration of 2×10^5 per mL in 8 mL, respectively.
3. Using an electronic multichannel pipette, plate the cell suspensions in a white plate at 50 μL per well and centrifuge the plate at 500 × *g* for 1 min.
4. Add compounds to be tested and TCIP1 as a positive control from 10 mM stock solutions in DMSO using a compound dispensing system and normalize to a final DMSO concentration of 0.1%.
5. Place the plate back in the incubator and incubate for 24 h.

3.3 Luciferase Assay Readout

1. Remove the plate from the incubator and place it at RT for 15 min (*see* **Note 5**).
2. Bring a 6 mL aliquot of the luciferase reagent to RT (*see* **Note 6**).
3. Using the electronic multichannel pipette, add 12.5 μL of luciferase reagent per well to the assay plate.
4. Incubate the plate for 10 min.
5. Read the luciferase signal using PheraSTAR FSX with the following settings (*see* **Note 7**):

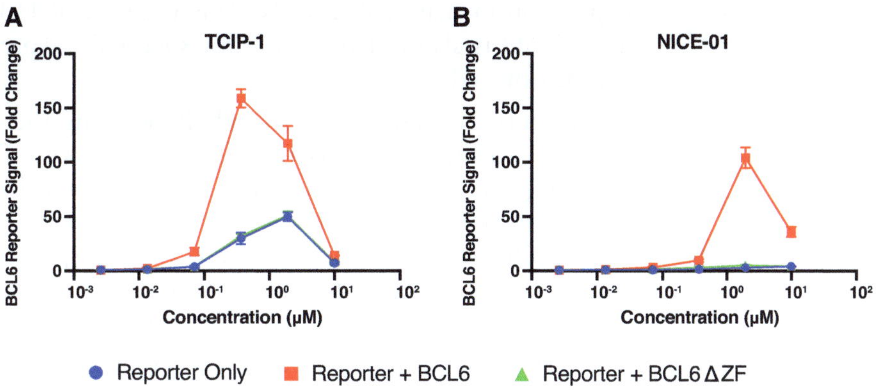

Fig. 4 Dose-response activation of the BCL6 transcriptional reporter by (**A**) TCIP1 and (**B**) NICE-01. Data represents the mean and standard deviation of triplicates, N = 3

a. Optic Module: LUM Plus.
b. Well Scan: None.
c. Optic: Top.
d. Settling Time: 0.5 s.
e. No. of Kinetic Windows: 1.
f. Number of cycles: 5.
g. Measurement start time: 0.0.
h. Measurement interval time: 1.0.
i. Cycle time: 617 s (See **Note 8**).

6. Analyze data by averaging each of the 5 cycle replicates to obtain the final signal for each well. Afterwards, normalize to the mean of the DMSO controls and plot against the dose-response of compound concentrations (*see* **Note 9**). Example data for dose-response of TCIP1 and NICE-01 are shown in Fig. 4.

4 Notes

1. Titration of compounds in DMSO can be added using various other dispensers (e.g., Echo Acoustic liquid handling instruments, Tecan Digital Microplate Dispenser D300e) or manually.
2. Compatible multi-well plate readers can be used, given they have the capability to measure luminescence.
3. The plasmid expressing the transcription factor can be prepared in any other backbone, allowing for transient transfection and

expression in mammalian cells. The addition of HiBiT, Flag, and dTAG is also not necessary for successful expression and can be omitted.

4. Cells can be collected enzymatically by washing the plate with 5 mL prewarmed DPBS followed by the addition of 1 mL per plate trypsin and 5 min incubation at 37°C. After detachment, gently tap the plate to release the cells and add 5 mL culture medium to neutralize the trypsin. The cell suspension is then transferred to a centrifuge tube, centrifuged (5 min at 300 × *g*), and the cell pellet resuspended in 5 mL medium.
5. Allow the plate to equilibrate to room temperature to avoid a temperature gradient, which may affect the uniformity of the luminescence readout.
6. For each new kit, 100 mL of Bio-Glo Luciferase Assay Buffer is added to Bio-Glo Substrate and can be aliquoted into 6 mL per 15 ml Falcon tube and frozen at −20 for further use (equivalent for 384-well plate).
7. These settings can be adjusted if needed and translated to other plate readers. The measurement interval time can be adjusted to another time and does not affect assay performance.
8. The cycle time is adjusted to the number of wells in the assay. Data can be analyzed using GraphPad Prism or similar software.
9. Data can be analyzed and plotted using GraphPad Prism or similar software.

Acknowledgments

We thank all members of the Nowak Lab and Institute of Structural Biology for useful discussions in the development of the assay. R.P.N. is a member of the excellence cluster ImmunoSensation3 funded by the Deutsche Forschungsgemeinschaft (DFG, German Research Foundation) under Germany's Excellence Strategy—EXC2151—390873048.

Conflict of Interest There are no conflicts.

References

1. Gourisankar S, Krokhotin A, Ji W, Liu X, Chang CY, Kim SH, Li Z, Wenderski W, Simanauskaite JM, Yang H, Vogel H, Zhang T, Green MR, Gray NS, Crabtree GR (2023) Rewiring cancer drivers to activate apoptosis. Nature 620(7973):417–425. https://doi.org/10.1038/s41586-023-06348-2
2. Fan F, Wood KV (2007) Bioluminescent assays for high-throughput screening. Assay Drug Dev Technol 5(1):127–136. https://doi.org/10.1089/adt.2006.053
3. Chiarella AM, Butler KV, Gryder BE, Lu D, Wang TA, Yu X, Pomella S, Khan J, Jin J,

Hathaway NA (2020) Dose-dependent activation of gene expression is achieved using CRISPR and small molecules that recruit endogenous chromatin machinery. Nat Biotechnol 38(1):50–55. https://doi.org/10.1038/s41587-019-0296-7

4. Sarott RC, Gourisankar S, Karim B, Nettles S, Yang H, Dwyer BG, Simanauskaite JM, Tse J, Abuzaid H, Krokhotin A, Zhang T, Hinshaw SM, Green MR, Crabtree GR, Gray NS (2024) Relocalizing transcriptional kinases to activate apoptosis. Science 386(6717):eadl5361. https://doi.org/10.1126/science.adl5361
5. Zhu X, Byun WS, Pieńkowska DE, Nguyen KT, Gerhartz J, Geng Q, Qiu T, Zhong J, Jiang Z, Wang M, Sarott RC, Hinshaw SM, Zhang T, Attardi LD, Nowak RP, Gray NS (2024) Activating p53^{Y220C} with a mutant-specific small molecule. bioRxiv:2024.2010.2023.619961. https://doi.org/10.1101/2024.10.23.619961
6. Nix MN, Gourisankar S, Sarott RC, Dwyer BG, Nettles SA, Martinez MM, Abuzaid H, Yang H, Wang Y, Simanauskaite JM, Romero BA, Jones HM, Krokhotin A, Lowensohn TN, Chen L, Low C, Davis MM, Fernandez D, Zhang T, Green MR, Hinshaw SM, Gray NS, Crabtree GR (2025) A bivalent molecular glue linking lysine acetyltransferases to oncogene-induced cell death. bioRxiv:2025.2003.2014.643404. https://doi.org/10.1101/2025.03.14.643404
7. Nabet B, Roberts JM, Buckley DL, Paulk J, Dastjerdi S, Yang A, Leggett AL, Erb MA, Lawlor MA, Souza A, Scott TG, Vittori S, Perry JA, Qi J, Winter GE, Wong KK, Gray NS, Bradner JE (2018) The dTAG system for immediate and target-specific protein degradation. Nat Chem Biol 14(5):431–441. https://doi.org/10.1038/s41589-018-0021-8
8. Nabet B, Ferguson FM, Seong BKA, Kuljanin M, Leggett AL, Mohardt ML, Robichaud A, Conway AS, Buckley DL, Mancias JD, Bradner JE, Stegmaier K, Gray NS (2020) Rapid and direct control of target protein levels with VHL-recruiting dTAG molecules. Nat Commun 11(1):4687. https://doi.org/10.1038/s41467-020-18377-w
9. Gibson WJ, Sadagopan A, Shoba VM, Choudhary A, Meyerson M, Schreiber SL (2023) Bifunctional small molecules that induce nuclear localization and targeted transcriptional regulation. J Am Chem Soc 145(48):26028–26037. https://doi.org/10.1021/jacs.3c06179

Part IV

Methods for Other Eukaryotic Organisms

Chapter 11

Studying Glycan-Dependent ERAD of Misfolded Glycoproteins in Plants

Ulrike Vavra, Christiane Veit and Richard Strasser

Abstract

The endoplasmic reticulum (ER) is the site where proteins that are synthesized and destined for the secretory pathway fold into their native conformations. Genetic mutations, oxidative stress, reduced glycosylation, and disruption to ER folding and quality control systems are all factors that have been demonstrated to impair this process, leading to the production of misfolded proteins. It is imperative for the survival of cells that these misfolded proteins are recognized by specific signals and directed towards degradation via ER-associated degradation (ERAD), ER-phagy, or related ER-to-lysosome/vacuole transport processes. A pivotal step in this process is the recognition of specific degrons and the distinction between terminally misfolded proteins that necessitate clearance and folding intermediates that have the potential to reach their final conformation. A significant proportion of proteins that traverse the secretory pathway in eukaryotes undergo N-glycosylation. The attached N-glycans not only facilitate the folding and quality control of glycoproteins, but can also be processed to display a glycan degron signal necessary to initiate ERAD of glycoproteins. Here, we describe methods for investigating glycan-related degrons that are key determinants for the ERAD of misfolded glycoproteins in plants.

Keywords N-glycosylation, Glycoprotein, Quality control, ERAD, Protein degradation, *Arabidopsis thaliana*

1 Introduction

Glycosylation is a widespread modification of newly synthesized proteins in the endoplasmic reticulum (ER). Depending on how the oligosaccharide is linked to the amino acid side chain of the protein, there are two main types of glycosylation: N- and O-glycosylation. Both types affect protein folding; this role is however more clearly established for N-glycosylation, whereby the N-glycan, which is attached to an asparagine in a specific sequence motif (Asn-X-Ser/Thr; X can be any amino acid except proline), is often attached co-translationally. Glycosylation can directly contribute to the folding process or facilitate interaction with specific lectins, such as calnexin

Maria W. Górna et al. (eds.), *Degrons*, Methods in Molecular Biology, vol. 3069,
https://doi.org/10.1007/978-1-0716-5508-5_11,

and calreticulin, which promote folding in association with interacting protein partners. [1, 2].

A hallmark of N-glycosylation is the *en bloc* transfer of a conserved preassembled oligosaccharide ($Glc_3Man_9GlcNAc_2$) (Fig. 1A) from the lipid carrier dolichol pyrophosphate to selected N-glycosylation sites, which is catalyzed by the oligosaccharyltransferase (OST) complex in all eukaryotes [3–5]. Upon transfer, the attached N-glycans are quickly trimmed by glucosidase I and II, which generates a monoglucosylated N-glycan that interacts with calnexin and calreticulin as part of the glycan-dependent ER-quality control [6]. Eventually, however, terminally misfolded glycoproteins are sensed, and the N-glycan is trimmed by specific α-mannosidases to expose a glycan degradation signal [7]. Notably, by exposing distinct glucose or mannose residues, the N-glycans determine the fate of

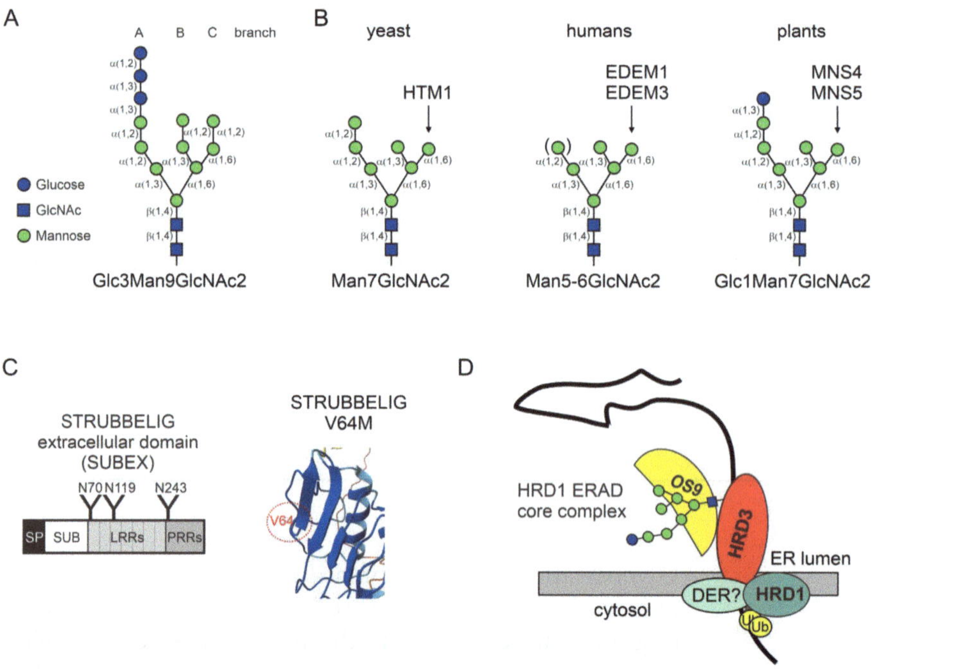

Fig. 1 (**A**) Schematic representation of the initially transferred oligosaccharide with the three branches. (**B**) Processed N-glycans exposing the α1,6-mannose residue on the C-branch. These N-glycans have been detected on glycoproteins subjected to ERAD in different species. The α-mannosidases (HTM1, EDEM1, EDEM3, MNS4, MNS5) responsible for the generation of the glycan degradation signal are indicated. (**C**) Illustration of the domain structure of the extracellular domain from Arabidopsis STRUBBELIG (UniProt: Q8RWZ1). The N-glycosylation sites are indicated. SP, signal peptide; SUB, SUB-domain; LRR, leucine-rich repeat; PRR, proline-rich repeat. A snapshot from the predicted STRUBBELIG structure is given to illustrate the location of the V64 amino acid that is mutated in SUBEX-V64M (Image was taken from the AlphaFold Protein Structure Database—AF-Q8RWZ1-F1-v4) [25]. (**D**) Schematic illustration of the plant HRD1 core complex consisting of the lectin OS9, the adaptor protein HRD3, the E3 ubiquitin ligase HRD1, and membrane-embedded DER proteins that are likely involved in retrotranslocation. The misfolded protein with the attached glycan is shown in black

glycoproteins in the ER. Glycan recognition and trimming steps are an essential part of the ER-quality control and glycan-dependent ERAD process, which is conserved in yeast, humans, and plants. Efficient clearance of misfolded glycoproteins from the ER lumen requires a bipartite signal composed of a specific N-glycan degron and a polypeptide degron.

1.1 The N-Glycan Degron

For persistently non-native glycoproteins, the initiation of ERAD requires the cleavage of mannose residues by specific α-mannosidases in the ER. In the context of mammalian cells, ER-degradation-enhancing α-mannosidase-like (EDEM) 2 cleaves a mannose residue from the B-branch (Fig. 1B). This is subsequently followed by mannose trimming on the C-branch, which is catalyzed by EDEM1/EDEM3 [8, 9]. In *Saccharomyces cerevisiae*, the mannosidase-like HTM1 enzyme removes the mannose from the C-branch [7]. In the model plant species *Arabidopsis thaliana*, two α-mannosidases (MNS4 and MNS5) are involved in this step [10]. In all these species, the trimming of mannose results in the formation of an exposed α 1,6-mannose residue on the C-branch (Fig. 1B). This glycan moiety is subsequently recognized by the ER-resident mannose-binding lectin YOS9 (yeast), OS9 or XTP3B (humans) or OS9 (plants). The lectin binding directs the misfolded glycoprotein to the membrane-embedded HRD1 E3 ubiquitin ligase complex for retrotranslocation, ubiquitination, and proteasomal degradation in the cytosol [11, 12]. The subsequent steps of the ERAD pathway have been extensively documented in yeast, human, and the model plant *Arabidopsis thaliana*. In contrast, the recognition step that initiates mannose trimming by the aforementioned enzymes remains less well understood and is hypothesized to involve the recognition of a non-native polypeptide conformation by α-mannosidases and associated proteins like protein disulfide isomerases [13, 14].

1.2 The Polypeptide Degron

The present methods will focus on polypeptides with lesions in a luminal domain (ERAD-L substrates). In the plant kingdom, the aforementioned lesions that are known to target glycoproteins to ERAD comprise the substitution of a single cysteine residue (SUBEX-C57Y; BRI1-5) [15, 16] and the presence of an additional hydrophobic amino acid in a conserved repeat or β-strand (SUBEX-V64M, BRI1-9) (Fig. 1C) [17, 18]. In properly folded secretory proteins, cysteines typically from disulfide bonds. Mutation of a single cysteine residue generates an unpaired cysteine, which destabilizes the protein. In the case of SUBEX-C57Y it has been demonstrated that the formation of the N-glycan degradation signal and the presence of the C57Y mutation are both prerequisites for HRD1 complex mediated ERAD (Fig. 1D) [15]. In this study, we present protocols designed to identify and characterize glycan-dependent ERAD substrates in plants.

2 Materials

1. *Arabidopsis thaliana* seeds with characterized defects in lipid-linked oligosaccharide assembly can be purchased from the Versailles Arabidopsis Stock Center (e.g., *alg12*–ID: FLAG_310A12; https://publiclines.versailles.inrae.fr/) [19]. Seeds with defects in ERAD are available from the European Arabidopsis Stock Center (*os9*–NASC ID: N529413; *hrd3*–NASC ID: N663635; *mns4 mns5*–double mutant can be obtained by crossing of NASC ID N656907 and NASC ID N162962, https://arabidopsis.info/BasicForm) [10].
2. *Nicotiana benthamiana* wild type or the glycoengineered ΔXT/FT line that is deficient in the formation of complex N-glycans with core α1,3-fucose [20] can be requested from the Strasser group.
3. Safe-lock microcentrifuge tubes.
4. Mixer mill with steel beads.
5. Container with liquid nitrogen.
6. Refrigerated bench-top centrifuge with a relative centrifugal force of 10.000 or higher and a rotor for 1.5/2 mL microcentrifuge tubes.
7. Thermo block.
8. Orbital shaker.
9. Vertical electrophoresis system.
10. Tank transfer system.
11. Power supply with at least 100 V and 400 mA.
12. SDS-PAGE (10%–12%) gels.
13. Tris/glycine/SDS-PAGE running buffer: 25 mM Tris, 192 mM glycine, 0.1% (w/v) SDS.
14. 3x Laemmli sample buffer: 187.5 mM Tris pH 6.8, 30% (w/v) glycerol, 6% (w/v) SDS, 15% (v/v) β-mercaptoethanol, 0.15% (w/v) bromophenol blue.
15. Pre-stained protein marker.
16. Tris/glycine/methanol transfer buffer: 25 mM Tris, 192 mM glycine, 20% (v/v) methanol.
17. Gel blotting paper.
18. Nitrocellulose blotting membrane.
19. 1× PBS: phosphate buffered saline: 137 mM NaCl, 2.7 mM KCl, 10 mM Na_2HPO_4, 2 mM KH_2PO_4, pH 7.4.
20. PBST: 1× PBS + 0.1% (v/v) Tween 20.
21. 1× TBS: Tris buffered saline (25 mM Tris, pH 7.5, 150 mM NaCl, 2 mM KCl).

22. TBST: 1× TBS + 0.1% (v/v) Tween 20.
23. Blocking solution for immunoblots: TBST + 5% (w/v) skimmed milk powder or PBST + 3% (w/v) BSA.
24. Western blotting detection reagent: for example, SuperSignal West PICO Plus or Agrisera ECL BRIGHT chemiluminescent substrates.
25. Mouse monoclonal GFP antibody (Roche: 11814460001). 1:2000 diluted in PBST + 3% (w/v) BSA.
26. Mouse monoclonal actin antibody (Sigma-Aldrich: A0480). 1:3000 diluted in PBST + 1% (w/v) BSA.
27. Mouse monoclonal α-tubulin antibody (Sigma-Aldrich: T6074). 1:5000 diluted in PBST + 1% (w/v) BSA.
28. Mouse monoclonal RFP antibody (Chromotek: 6g6). 1:2000 diluted in PBST + 0.5% (w/v) BSA.
29. Anti-mouse IgG-peroxidase (such as Sigma A9044): 1:10000 diluted in TBST for detection of GFP, actin, tubulin, and RFP.
30. Endoglycosidase H.
31. Empty microspin chromatography columns.
32. Kifunensine: class I α-mannosidase inhibitor, dissolved in ultrapure water.
33. MS medium: 1/2 × Murashige & Skoog (MS) medium including MES buffer supplemented with 0.8% (w/v) agar and 1% (w/v) sucrose.
34. MS liquid medium: 1/2 × MS medium including MES buffer supplemented with 1% (w/v) sucrose.
35. Ponceau S in 5% (v/v) acetic acid.
36. Chemiluminescence imaging system.
37. Flatbed scanner.
38. Cycloheximide: dissolved in dimethyl sulfoxide.
39. 6- well cell culture multiwell plates.
40. Infiltration buffer: 50 mM MES, pH 5.6, 2 mM Na_3PO_4 $12H_2O$, 5 mg/mL D-glucose, 0.1 mM acetosyringone.
41. OD_{600} photometer and cuvettes.

3 Methods

3.1 Monitoring of N-Glycan Dependent Protein Degradation in Arabidopsis Seedlings

1. Respective *A. thaliana* seeds were germinated on 1/2 × MS agar plates supplemented with 1% (w/v) sucrose. Plates were incubated for 8–12 days (*see* **Note 1**) in a dedicated growth chamber at 21°C under controlled light (16 h light/8 h dark) conditions.

2. 8–12-days-old *A. thaliana* seedlings were detached from the plates and submerged in 6- well plates filled with MS liquid medium supplemented with 20 μM kifunensine or as a control in liquid MS medium without kifunensine.
3. Incubate at 21°C under controlled light (16 h light/8 h dark) conditions with gentle shaking on an orbital shaker for a duration of 24 h.
4. Harvest 10–100 mg of plant material (e.g., seedlings) from the mutants and *A. thaliana* wild type, then transfer the plant material to 2 mL safe-lock microcentrifuge tubes containing two 5 mm diameter steel beads per tube.
5. Submerge the 2 mL tubes containing the plant material in a container of liquid nitrogen. Remove the tubes after 3 min, place them in a mixer mill and run the mixer mill for 2 min at an amplitude of 50–60. Add 4 μL of PBST extraction buffer for every mg of plant material. Mix briefly and transfer the mixture to a 1.5 mL tube. Incubate on ice for 15 min, inverting the tube every 3 min.
6. Centrifuge twice for 15 min at 9600 ×*g* at 4°C, transferring the supernatant to a new tube each time. Mix the samples with 3× Laemmli sample buffer and heat to 95°C for 5 min. Load the SDS-PAGE gel with 20 μL (*see* **Note 2**) and run the protein extract for 1.5 h at 100 V. In order to monitor the accumulation of the ERAD substrate, immunoblotting should be performed with appropriate antibodies (e.g., GFP antibody for ERAD substrates fused to GFP).
7. For a quantitative approach, the same membrane should be probed with an antibody directed against an unrelated control protein (e.g., actin or tubulin), and the membrane should be stained with Ponceau S solution to visualize the total protein in each lane (*see* **Note 3**). Detection of the signals on the membrane is done using a chemiluminescence imaging system and Ponceau S-stained membranes are scanned using a flatbed scanner.
8. An increased signal of the ERAD substrate protein in the protein extracts of seedlings that were incubated with kifunensine indicates the involvement of a glycan-dependent degradation pathway that depends on trimming of mannose residues by class I α-mannosidases (*see* **Note 4**) (Fig. 2A, B).
9. To substantiate this finding, the misfolded glycoprotein can be stably expressed in different genetic mutants (e.g., *mns4 mns5, alg12, and os9*), which have distinct defects in glycan-dependent ERAD [15]. In line with blocked glycan-dependent ERAD there is no increase of the misfolded glycoprotein in the presence of kifunensine in these mutants (Fig. 2A, B).

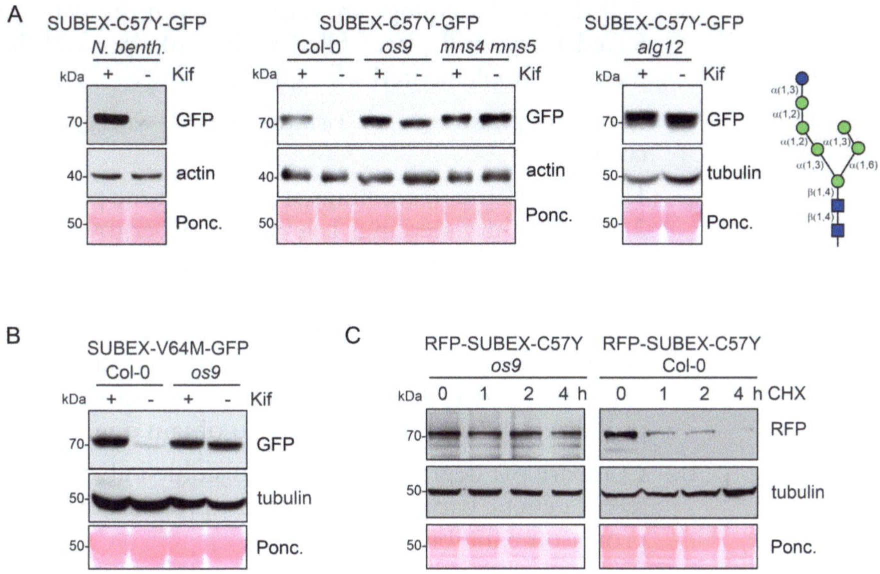

Fig. 2 (**A**) Immunoblot analysis of the ERAD substrate SUBEX-C57Y-GFP. Leaves of infiltrated *N. benthamiana* or Arabidopsis seedlings from Col-0 wild type, *os9*, *mns4 mns5* or *alg12* ERAD mutants stably expressing SUBEX-C57Y-GFP were incubated with/without kifunensine (Kif). Protein extracts were subjected to SDS-PAGE and immunoblotting. Detection of endogenous actin or α-tubulin serves as a control. The proposed N-glycan structure present on misfolded glycoproteins in the *alg12* mutant is shown in the illustration. The protein is not subjected to ERAD because it lacks the free α1,6-mannose. (**B**) Immunoblot analysis of the ERAD substrate SUBEX-V64M-GFP stably expressed in Col-0 or *os9*. (**C**) Cycloheximide (CHX)-chase assay. Seedlings were incubated for the indicated time points in CHX and subsequently subjected to immunoblotting

10. A fast approach to assess the impact of kifunensine on an expressed protein is the transient expression of the glycoprotein by agroinfiltration of *N. benthamiana* leaves. An overnight culture of Agrobacteria carrying an expression vector for the misfolded glycoprotein is harvested by low-speed centrifugation. The Agrobacteria are resuspended in infiltration buffer (final $OD_{600} = 0.1$ to 0.3) and infiltrated with a needle-free syringe into the abaxial side of 5-week-old leaves. After 24 h, 20 μM kifunensine in infiltration buffer without acetosyringone is infiltrated into the same leaf area and after another 24 h the infiltrated leaves are detached from the plants. Homogenization, protein extraction, and immunoblotting are carried out as described for *A. thaliana* (Fig. 2A).

3.2 CHX Treatment of Seedlings

1. *A. thaliana* seedlings are grown as described in Sect. 3.1.
2. Prepare a cycloheximide (CHX) working solution (final concentration 100 μg/mL) by dilution of a 100 mg/mL stock in liquid MS medium.

3. 7–9-days-old seedlings are removed from the plate and transferred to 6- well plates filled with 6 mL/per well liquid MS medium supplemented with CHX.
4. Incubate plates on an orbital shaker with gentle shaking (60 rpm).
5. Harvest 200 mg of seedlings at different time points (e.g., 0, 1, 2, and 4 h), dry seedlings briefly on a paper towel, and transfer the seedlings into 2 mL safe-lock microcentrifuge tubes with steel beads. Freeze in liquid nitrogen, extract and analyze the proteins as described in Sect. 3.1. Analyze the abundance of an unrelated protein (e.g., actin or tubulin) at all time points (Fig. 2C) and normalize the protein signal of the misfolded protein to the unrelated protein to obtain quantitative data.
6. Compare the degradation of the misfolded glycoprotein in different genetic backgrounds or in the presence of kifunensine or proteasome inhibitors.

3.3 Characterization of ER-Type Oligomannosidic N-Glycans

It is a well-established fact that ER-retained proteins typically bear oligomannosidic N-glycans. This phenomenon occurs because these proteins do not come into contact with the Golgi-localized machinery responsible for processing of complex N-glycans [6]. Oligomannosidic N-glycans are susceptible to digestion by endoglycosidase H (Endo H), an enzyme that cleaves the oligosaccharide at the chitobiose core. Consequently, digestion of oligomannosidic N-glycans by Endo H will result in a reduction in molecular weight, which can be monitored by SDS-PAGE and immunoblotting (Fig. 3A). By contrast, proteins that traffic through the Golgi apparatus usually carry complex N-glycans, which are resistant to Endo H digestion and do not display a similar change in molecular weight. N-glycans processed in the Golgi apparatus in mammals can be cleaved off by PNGase F digestion. However, this enzyme is blocked by the core α1,3-fucose commonly found on plant complex N-glycans [21, 22] (*see* **Note 5**).

1. Incubate 22.5 μL of the protein extract (1–20 μg of protein) with 2.5 μL 10× Glycoprotein denaturing buffer for 10 min at 95°C, transfer to ice and cool for 5 min (*see* **Note 6**).
2. Mix 22.5 μL of the denatured glycoproteins with 3 μL 10× Glycobuffer 3 (NEB), 3 μL ultrapure water, and 1.5 μL Endo H (*see* **Note 7**). For the control reaction, replace Endo H with ultrapure water.
3. Incubate for 120 min at 37°C and stop the reaction by heating to 95°C for 5 min.
4. Load samples on the SDS-PAGE gel and perform immunoblotting, for example, with a GFP antibody to detect the ERAD substrate protein fused to GFP.

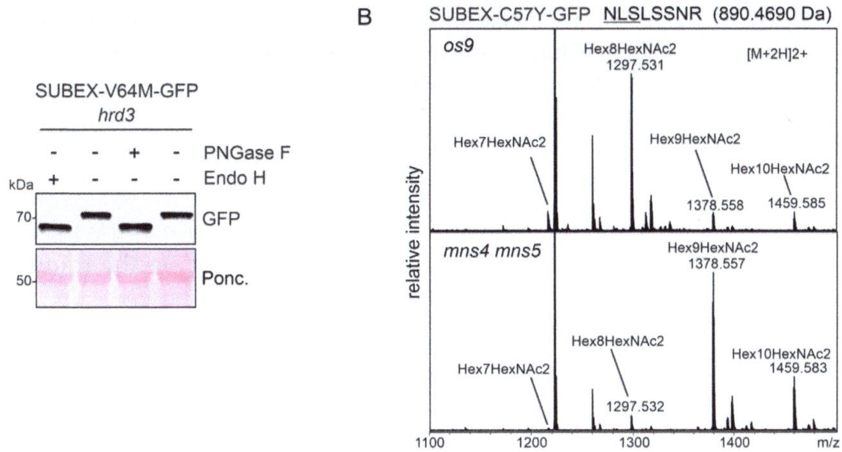

Fig. 3 (**A**) Endoglycosidase (Endo H or PNGase F) digestion of protein extracts from the Arabidopsis *hrd3* ERAD mutant stably expressing SUBEX-V64M-GFP. The shift in mobility following Endo H digestion indicates the presence of oligomannosidic N-glycans, which are characteristic of ER-retained proteins. (**B**) LC-ESI-MS analysis of the SUBEX-C57Y glycopeptide carrying the N-glycosylation site Asn119. SUBEX-C57Y-GFP was purified from the indicated Arabidopsis mutants. The N-glycan from the *mns4 mns5* mutant carries an additional hexose residue, which is consistent with the absence of mannose trimming activity that leads to the formation of the exposed α1,6-mannose residue on the C-branch (as shown in the *os9* mutant)

5. To confirm the presence of the oligomannosidic N-glycans and the exposed α1,6-mannose residue, the misfolded glycoprotein can be purified from total soluble proteins by affinity chromatography using (e.g., a GFP-nanobody coupled to agarose beads is commonly used to purify GFP-tagged proteins [15]). The purified glycoprotein can be proteolytically digested and glycopeptides subjected to LC-ESI-MS analysis to obtain the site-specific glycan composition (Fig. 3B) [15, 23] (*see* **Note 8**).

4 Notes

1. The seedlings can also be younger or slightly older, and similar protocols are available for detached leaves from plants grown on soil [11]. Seedlings older than 10 days typically have longer roots that can be recalcitrant to fully submerging in the 6-well plates filled with MS liquid medium.
2. Loading the same volume here usually produces highly comparable total protein amounts. Therefore, it is normally not necessary to measure the protein concentration prior to SDS-PAGE.
3. Bands can be quantified and the signal intensity can be normalized by the signal intensity of the detected control protein

or Ponceau S. Be aware that the linear range for quantification of bands is typically quite narrow. Densitometric analysis can be performed using a specific software provided by the used imaging system like the Fusion EvolutionCapt software (Vilber) or the Fiji open-source image processing package based on Image J.

4. Quantitative RT-PCR (RT-qPCR) may be performed to rule out that kifunensine has any effect on the transcription of the ERAD substrate.
5. The expression of glycoproteins in plants deficient in core α1,3-fucose formation has been reported, and can be used to investigate complex N-glycan presence via PNGase F digestion [20, 22].
6. Denaturation of the glycoproteins is necessary to facilitate the access of the endoglycosidase to all N-glycans.
7. Oligomannosidic N-glycans have been identified on secretory proteins that do not undergo trafficking through the Golgi, for instance, as a consequence of direct ER-to-vacuole protein trafficking. Furthermore, some N-glycans are not accessible for Golgi-mediated N-glycan processing (e.g., steric hindrance of N-glycan processing), resulting in the presence of oligomannosidic N-glycans on secreted or plasma membrane-anchored proteins. Consequently, further experimentation is required, such as confocal microscopy of fluorescent fusion proteins, to confirm the ER localization.
8. To distinguish between different isobaric mannosidic N-glycans, in vitro digestion with specific α1,2-, α1,3-, or α 1,6-mannosidases might be required prior to MS analysis. Alternatively, released N-glycans can be subjected to porous graphitic carbon (PGC)-liquid chromatography (LC) in conjunction with MS analysis [24].

Acknowledgments

We would like to thank Yun-Ji Shin (BOKU University) for help with immunoblots. We thank Daniel Maresch (BOKU Core Facility Mass Spectrometry) for conducting MS experiments. The MS equipment was kindly provided by the BOKU Core Facility Mass Spectrometry. This work was funded by the Austrian Science Fund (FWF) P35621 (Grant-DOI 10.55776/P35621).

References

1. Guay KP, Chou WC, Canniff NP, Paul KB, Hebert DN (2025) N-glycan-dependent protein maturation and quality control in the ER. Nat Rev Mol Cell Biol 26:926–939. https://doi.org/10.1038/s41580-025-00855-y
2. Kozlov G, Gehring K (2020) Calnexin cycle - structural features of the ER chaperone system. FEBS J 287(20):4322–4340. https://doi.org/10.1111/febs.15330
3. Braunger K et al (2018) Structural basis for coupling protein transport and N-glycosylation at the mammalian endoplasmic reticulum. Science 360(6385):215–219. https://doi.org/10.1126/science.aar7899
4. Jeong IS et al (2018) Purification and characterization of *Arabidopsis thaliana* oligosaccharyltransferase complexes from the native host: a protein super-expression system for structural studies. Plant J 94(1):131–145. https://doi.org/10.1111/tpj.13847
5. Wild R et al (2018) Structure of the yeast oligosaccharyltransferase complex gives insight into eukaryotic N-glycosylation. Science 359(6375):545–550. https://doi.org/10.1126/science.aar5140
6. Strasser R (2016) Plant protein glycosylation. Glycobiology 26(9):926–939. https://doi.org/10.1093/glycob/cww023
7. Clerc S et al (2009) Htm1 protein generates the N-glycan signal for glycoprotein degradation in the endoplasmic reticulum. J Cell Biol 184(1):159–172
8. Ninagawa S et al (2014) EDEM2 initiates mammalian glycoprotein ERAD by catalyzing the first mannose trimming step. J Cell Biol 206(3):347–356. https://doi.org/10.1083/jcb.201404075
9. George G et al (2020) EDEM2 stably disulfide-bonded to TXNDC11 catalyzes the first mannose trimming step in mammalian glycoprotein ERAD. Elife 9:e53455. https://doi.org/10.7554/eLife.53455
10. Hüttner S et al (2014) Arabidopsis Class I α -Mannosidases MNS4 and MNS5 are involved in endoplasmic reticulum-associated degradation of misfolded glycoproteins. Plant Cell 26(4):1712–1728. https://doi.org/10.1105/tpc.114.123216
11. Schoberer J, Vavra U, Shin YJ, Grunwald-Gruber C, Strasser R (2025) Elucidation of the late steps in the glycan-dependent ERAD of soluble misfolded glycoproteins. Plant J 121(1):e17185. https://doi.org/10.1111/tpj.17185
12. Wang HH, Biunno I, Sun S, Qi L (2025) SEL1L-HRD1-mediated ERAD in mammals. Nat Cell Biol 27(7):1063–1073. https://doi.org/10.1038/s41556-025-01690-1
13. Shenkman M et al (2018) Mannosidase activity of EDEM1 and EDEM2 depends on an unfolded state of their glycoprotein substrates. Commun Biol 1:172. https://doi.org/10.1038/s42003-018-0174-8
14. Zhao D, Wu X, Rapoport TA (2025) Initiation of ERAD by the bifunctional complex of Mnl1/Htm1 mannosidase and protein disulfide isomerase. Nat Struct Mol Biol 32(6):1006–1018. https://doi.org/10.1038/s41594-025-01491-y
15. Hüttner S et al (2014) A context-independent N-glycan signal targets the misfolded extracellular domain of Arabidopsis STRUBBELIG to endoplasmic-reticulum-associated degradation. Biochem J 464(3):401–411. https://doi.org/10.1042/BJ20141057
16. Hong Z, Jin H, Tzfira T, Li J (2008) Multiple mechanism-mediated retention of a defective brassinosteroid receptor in the endoplasmic reticulum of Arabidopsis. Plant Cell 20(12):3418–3429
17. Vaddepalli P, Fulton L, Batoux M, Yadav RK, Schneitz K (2011) Structure-function analysis of STRUBBELIG, an Arabidopsis atypical receptor-like kinase involved in tissue morphogenesis. PLoS One 6(5):e19730. https://doi.org/10.1371/journal.pone.0019730
18. Jin H, Yan Z, Nam K, Li J (2007) Allele-specific suppression of a defective brassinosteroid receptor reveals a physiological role of UGGT in ER quality control. Mol Cell 26(6):821–830
19. Veit C, König J, Altmann F, Strasser R (2018) Processing of the terminal Alpha-1,2-linked mannose residues from oligomannosidic. Front Plant Sci 9:1807. https://doi.org/10.3389/fpls.2018.01807
20. Strasser R et al (2008) Generation of glyco-engineered Nicotiana benthamiana for the production of monoclonal antibodies with a homogeneous human-like N-glycan structure. Plant Biotechnol J 6(4):392–402
21. Tretter V, Altmann F, März L (1991) Peptide-N4-(N-acetyl-beta-glucosaminyl)asparagine amidase F cannot release glycans with fucose attached alpha 1-3 to the asparagine-linked N-acetylglucosamine residue. Eur J Biochem 199(3):647–652
22. Strasser R, Altmann F, Mach L, Glössl J, Steinkellner H (2004) Generation of Arabidopsis thaliana plants with complex N-glycans lacking beta1,2-linked xylose and core alpha1,3-linked fucose. FEBS Lett 561(1–3):132–136
23. Stadlmann J, Pabst M, Kolarich D, Kunert R, Altmann F (2008) Analysis of immunoglobulin glycosylation by LC-ESI-MS of glycopeptides and oligosaccharides. Proteomics 8(14):2858–2871

24. Pabst M et al (2012) Isomeric analysis of oligomannosidic N-glycans and their dolichol-linked precursors. Glycobiology 22(3):389–399. https://doi.org/10.1093/glycob/cwr138

25. Jumper J et al (2021) Highly accurate protein structure prediction with AlphaFold. Nature 596(7873):583–589. https://doi.org/10.1038/s41586-021-03819-2

Chapter 12

Protein Depletion in *Caenorhabditis elegans* Using the Auxin-Inducible Degradation System

Jordan D. Ward

Abstract

The auxin-inducible degradation (AID) system is a powerful tool in modern molecular genetics that allows for conditional protein depletion. In *Caenorhabditis elegans*, tools exist to allow for rapid, tissue-specific depletion of target proteins. The system requires tagging a gene of interest with an AID degron and a transgene expressing the *Arabidopsis thaliana* TIR1 F-box protein, which can form a functional SCF ubiquitin ligase with endogenous Skp and Cullin proteins. Here, I describe how to generate degron knock-ins by CRISPR, cross them to TIR1-expressing strains, and perform depletion experiments. I provide a description of our current methods and highlight alternative approaches.

Keywords Auxin-inducible degradation system, AID, *C. elegans*, TIR1, CRISPR, Conditional protein depletion

1 Introduction

The auxin-inducible degron system harnesses a plant mechanism used to destroy IAA transcriptional regulators [1, 2]. The IAA proteins contain a degron that is recognized by a Skp1-Cullin-F-box (SCF) E3 ubiquitin ligase only in the presence of the plant hormone auxin (indole-3-acetic acid) [1, 3]. The F-box protein TIR1 provides substrate specificity, binding auxin and the degron in the IAA protein [3, 4]. The system works in a broad range of organisms and cell types due to the conserved nature of SCF ubiquitin ligases, in which the TIR1 F-box protein can form a complex with endogenous Skp and cullin proteins [5–14].

In *Caenorhabditis elegans*, AID tags are added to target proteins through CRISPR/Cas9-mediated genome editing. The most commonly used degron is a 44-amino acid sequence from IAA17 (AID*) [5], though alternate degrons (mAID, mIAA7) exist [15, 16]. Notably, mIAA7 has been shown to provide more robust depletion for several proteins, though fewer reagents currently exist to utilize it

Maria W. Górna et al. (eds.), *Degrons*, Methods in Molecular Biology, vol. 3069,
https://doi.org/10.1007/978-1-0716-5508-5_12,

[15]. A number of TIR1 strains are available with different pros and cons [17]. We currently use either a *promoter::TIR1::mRuby::unc-54 3'UTR* or *promoter::TIR1::F2A::mTagBFP2::AID*::NLS::tbb-2 3'UTR* transgene [5, 9]. The former provides faster depletion, and the latter allows for imaging of red fluorescent proteins and is useful for suppressor screens of AID-depletion phenotypes [9]. For some targets, there can be variable auxin-independent degradation that can be avoided by using a mutant F79G or F79A TIR1 transgene and a modified auxin (5-Phenyl-IAA; 5-Ph-IAA) [18–21].

In this methods chapter, I will focus on our most used approach using AID*, a *promoter::TIR1::F2A::mTagBFP2::AID*::NLS::tbb-2 3'UTR* transgene and either an IAA or K-NAA ligand. I direct readers to Ward et al. 2026, a review written in collaboration with the Caenorhabditis Genetics Center, launching a curated collection of protein depletion strains. The review highlights how to find strains and the pros and cons of different TIR1 strains, degron sequences, and auxin ligands. Note that when “I” is used, it denotes personal opinion while “we” indicates standard practice in my lab.

2 Materials

2.1 Bacterial Culture

1. 50 mg/mL streptomycin stock solution.
2. LB liquid media.
3. LB+Strep agar plate (*see* **Note 1**).
4. OP50 *E. coli* from the Caenorhabditis Genetics Center. We recommend using OP50-1, which is streptomycin-resistant and helps minimize contamination of the stock. https://cgc.umn.edu/strain/OP50-1

2.2 MYOB (Modified Youngren's, Only Bacto-Peptone) Plates (See Note 2)

1. PourBoy4 (Tritech Research Inc.) or a similar peristaltic pump for plate pouring.
2. Tris–HCl.
3. Tris base.
4. Bacto Peptone.
5. Cholesterol.
6. NaCl.

2.3 Cas9, trcRNA, and crRNA

1. We use IDT Alt-R™ tracrRNA (100 nmol), Alt-R™ crRNA (2 nmol), and Alt-R™ S.p. Cas9 Nuclease V3, 500 μg. We set up the reactions in 0.2 μL PCR tubes or PCR strips.
2. After receiving a fresh tube of trcRNA, add IDTE pH 7.5 to 3000–4000 ng/μL and confirm the concentration by Nanodrop. Cas9 comes as a glycerol stock at 10,000 ng/μL.

We store both trcRNA and Cas9 stocks at −80°C and make 30–50 µL working stocks that can be used and then stored at −20°C for several weeks. trcRNA is diluted to 400 ng/µL in IDTE, and Cas9 is diluted to 2500 ng/µL in 1× PBS.

3. Add 75–80 µL IDTE pH 7.5 to a new 2 nmol aliquot of crRNA and pipette up and down to mix. This should result in a concentration of ~200 ng/µL, which will allow for the addition of ~1 µL crRNA per reaction. Use the Nanodrop to confirm concentration and store at −80°C.

2.4 TIR1 Strains and AID-Tagged Alleles

Many TIR1 and AID-tagged alleles are available at the CGC. For guidance in strain selection, see the review on curated protein depletion strains [17]. One can cross existing AID-tagged alleles into tissue-specific or pan-somatic TIR1 lines. Sects. 3.4–3.10 outline how to generate an AID*::3xFLAG knock-in into a new gene. One can either do the CRISPR/Cas9-mediated genome editing directly in a TIR1 strain or generate the knock-in and cross it into a set of TIR1 lines. We recommend using control strains KRY88 and/or JDW92 when establishing the system in a new laboratory (*see* **Note 3**).
https://cgc.umn.edu/strain/KRY88
https://cgc.umn.edu/strain/JDW92

2.5 10× Phusion Buffer

Phusion Buffer (1×) = 20 mM Tris pH 8.5, 80 mM KCl; 1.5 mM Mg (Cl_2 or SO_4); 0.1% Tween 20
136 mL 1 M Tris Base
64 mL 1 M Tris–HCl
400 mL 2 M KCl
15 mL 1 M $MgCl_2$ or SO_4
100 mL 10% Tween 20 or 10 mL 100% Tween 20
dH_2O to 1 L
Mix and then sterilize using a Millipore Stericup or similar vacuum filtration system. Make 1 mL and 50 mL aliquots in sterile tubes and store at −20°C.

2.6 Platinum Worm Pick

See dx.doi.org/10.17504/protocols.io.b2p3qdqn for an appropriate protocol to make a worm pick. We use 0.5 mm platinum (Thermo).

2.7 Molecular Biology Reagents

a. Nuclease-free water (not DEPC treated).
b. 5× Phusion Green HF Reaction Buffer (Thermo).
c. dNTPs (e.g., NEB).
d. Phusion (homemade; can also purchase from Thermo Scientific or UC Berkeley Macrolabs (https://qb3.berkeley.edu/facility/qb3-macrolab/#facility-rates)).
e. AMPureXP beads (Beckman Coulter).
f. Magnetic rack (e.g., NEB).

3 Methods

3.1 Make OP50 Stock for Seeding Plates

1. Streak out a glycerol stock of OP50 on an LB+streptomycin agar plate and incubate overnight (~16 h) at 37°C. Plates can be stored at 4°C for up to 1 month.
2. Make an OP50 culture. In a 2–4 L flask that has been autoclaved and covered with foil, add:
   a. 1 L LB.
   b. 1 mL streptomycin (50 mg/mL; 1000×).
   c. One OP50-1 colony, picked with a sterile pipette tip and dropped into the flask.

   Incubate overnight (~16 h) at 37°C with shaking at 225 rpm.
3. The next day, divide the culture into two 500 mL sterile bottles. Weigh the bottles first to make sure they are balanced, then adjust by transferring culture from the heavier bottle to the lighter one using a sterile 25 mL pipette. Centrifuge the bottles at 3220 rcf in a swinging-bucket centrifuge. Carefully pour off the supernatant without disturbing the bacterial pellet. Add 100 mL of LB+Strep to one bottle, then swirl and vortex to fully resuspend the pellet. Pour this resuspension into the second bottle and resuspend that pellet in the same way. Finally, transfer the OP50 culture into three 50 mL conical tubes, labeling them sequentially (e.g., 1, 2, 3) to track the lot.

3.2 MYOB (Modified Youngren's, Only Bacto-Peptone) Plates (See Note 2)

1. Preparation of MYOB powder (sufficient for 100 liters of media): Weigh and combine the following in a one-gallon Ziplock freezer bag:
   a. 55 g Tris–HCl.
   b. 24 g Tris base.
   c. 310 g Bacto Peptone.
   d. 800 mg Cholesterol.
   e. 200 g NaCl.

   Mix thoroughly and store at −20°C for up to 6 months.
2. Preparation of agar media for plates:
   Add 5.9 g of MYOB powder and 20 g of agar to 1 L of Milli-Q water in a 4 L Erlenmeyer flask. For 2 L and 4 L volumes of media, scale accordingly (*see* **Note 4**). Mix well with a magnetic stir bar (leave the stir bar in the flask). Cover the flask with aluminum foil and secure with autoclave tape.
3. Prepare plate pouring accessory materials: wrap the two tubes for the PourBoy dispenser in a foil packet, and prepare two 1 L

bottles with Milli-Q water (1 L) covered with foil, not a lid; if pouring 10 cm or 15 cm Petri dishes, include a 1 L Nalgene beaker with a pouring spout and cover it with foil. For larger Petri dishes, we hand-pour plates using the Nalgene beaker. Add autoclave tape to the foil packet, two 1 L bottles with millQ water, and the optional 1 L Nalgene beaker. Autoclave for 55 min on liquid cycle.

4. Put MYOB media on a magnetic stirrer and turn it on to a medium speed to mix the media. Cool MYOB down to 55°C. Stir at a speed that mixes media without producing bubbles. We use an Etekcity Infrared Thermometer to monitor temperature. While cooling, prepare auxin.
5. Auxin preparation and addition.
   Dissolve auxin in ethanol, then add to liquid media. To make 1 mM auxin plates, dissolve 175 mg of auxin in 2.5 mL of 100% ethanol and add to 1 L of MYOB agar media. To make 2 L and 4 L of media, scale up accordingly. To make 4 mM auxin plates, add 2.5 mL of a 1.6 M stock to 1 L of MYOB agar media. To make 2 L and 4 L of media, scale up accordingly. (For 1.6 M stock preparation, *see* **Note 5**.)
6. Preparation of matched 0.25% ethanol control plates:
   Add 2.5 mL of 100% ethanol per 1 L. For 2 L and 4 L media volumes, scale the ethanol accordingly (5 mL and 10 mL ethanol, respectively).

3.3 Prepare PourBoy and Pour Plates (See Note 6)

3.3.1 Preparation

1. Clean area. Spray the work area with 70% ethanol and wipe down. Set up a Bunsen burner near the pouring area. Place a bottle of 3% peroxide near the PourBoy. Arrange sterile Petri dishes in stacks, slightly fanned for easy access.
2. Clean hands. Spray hands with 70% ethanol.

3.3.2 Set-Up and Sterilization of Tubing

1. Open tubing. Carefully open the autoclaved foil packet containing tubing, being careful not to touch the tube ends to avoid contamination.
2. Connect the first tubing to PourBoy4. Take the forked end of the tubing approximately 1 inch from the tip and connect it to the right side of the PourBoy (wider port; this is the dispensing line). Place the other end of this tubing into the peroxide bottle (*see* **Note 7**).
3. Attach the second tubing to PourBoy4. Take the blunt end of the other tube ~1 inch from the tip and connect it to the **left side of the PourBoy**. Place the unattached end into peroxide.

4. Sterilize system. Press the "mode" button until the machine is in Auto Mode. Press the foot pedal to begin continuous cycling of peroxide. Run for **at least 10 min**.

3.3.3 Running Media

1. Check PourBoy4. The machine should make a buzzing sound when empty and a smooth purring noise when running properly.
2. Prepare agar media. Allow agar to cool to **55–65°C** using an infrared thermometer gun to monitor temperature (see Step **3.2.4**). If preparing auxin or ethanol plates, add supplements once the medium reaches this temperature.
3. Heat sterilization. Transfer both tubing ends from peroxide into an autoclaved Milli-Q water bottle. Secure tubes with the autoclaved foil cap to keep them sterile. Press the foot pedal to begin continuous cycling of autoclaved water. Let the water run until the tubing feels hot (heat sterilization step).
4. Transfer to media. Move the free end of the **left tube** into your sterile media container, being careful not to contaminate the tube ends. Set the PourBoy to the desired dispense volume.

3.3.4 Pouring Plates

1. Sanitize hands again. Spray hands with 70% ethanol.
2. Set PourBoy to dispense the desired amount of media. We use:
   a. 10 mL MYOB media for 6 cm plates.
   b. 5 mL MYOB media per well of a 6-well plate.
   c. 2.5 mL MYOB media per well of a 12-well plate.
   d. 1 mL MYOB media per well of a 24-well plate.**

   **Can also transfer to a sterile 1 L beaker and use a repeat pipette to dispense into the wells.
3. Dispense agar/media. Grab the **output tube** (from the right side of PourBoy) about 1 in. from the end. Dispense a small amount into a waste container to eliminate bubbles and ensure smooth flow. Lift the lid of the Petri dish or plate. Position tubing over plate/well. Press the foot pedal to dispense. Optional: use a kitchen butane torch to remove bubbles from the poured medium.

3.3.5 Clean Up

1. Rinse system. Place both tubing ends into unused autoclaved Milli-Q water. Run continuously for 3 min to flush the machine and tubing.
2. Dry system. Remove tubing ends. Allow the system to run dry.
3. Storage. Disconnect tubing. Allow to dry completely, then put away.

3.3.6 Storage and Seeding of Plates

1. Dry plates. Leave plates agar side up overnight. The following day, flip them upside down (agar facing down) and use Kimwipes to remove any condensation on the lids. Leave plates to dry on the bench for 1–3 days.
2. Seed plates. Use a repeat pipette to aliquot OP50 onto the plate. For 6-well plates, pipette 100 μL; for 6 cm plates, pipette 100 μL and spread with a disposable plastic spreader; for 10 cm plates, pipette 300 μL and spread with a disposable plastic spreader. Leave plates with the agar surface facing up at room temperature until the bacterial lawn dries.
3. Storage. Flip plates and store at 4°C for up to a month.

3.4 Background Research to Design AID Knock-Ins

Goal: For the AID system to work, the degron must be exposed to either the nucleus or cytosol.

1. Go to www.wormbase.org and enter the name of the gene of interest. Review available information on predicted subcellular localization and function. Determine whether the protein is predicted to be secreted, localized to the endocytic pathway, or membrane-bound. Check for the number of isoforms and decide whether to tag all isoforms or only a specific one. Identify which domains, if any, are predicted to be cytosolic. Our default strategy is C-terminal tagging, as the presence of introns in the gene can promote robust expression and reduce the risk of germline silencing. This preliminary work will guide design, which may be unique for each gene.
2. If no information is available, run the protein sequence through a signal peptide prediction tool such as https://services.healthtech.dtu.dk/service.php?SignalP-5.0
3. From WormBase, open the Sequences tab and click the protein name (not the double-helix logo). This will display the predicted domains of the protein. Avoid inserting fluorescent proteins or other large tags adjacent to functional domains, as such an insertion could disrupt protein activity. Additionally, run the gene of interest through GExplore to assess expression level and tissue specificity. This dataset is derived from L2 larvae and may not capture stage-restricted expression patterns. http://genome.sfu.ca/gexplore/gexplore_search_tissues.html

3.5 Selecting a crRNA Site

1. Make a sequence file for the genomic DNA of your gene of interest. We use "A Plasmid Editor" because it is free and easy to use (Fig. 1B).
2. Go to wormbase.org and search for the gene you wish to tag. Based on the background work in 3.1, go to the *Sequence* tab on Wormbase and retrieve the genomic sequence with 1000 bp

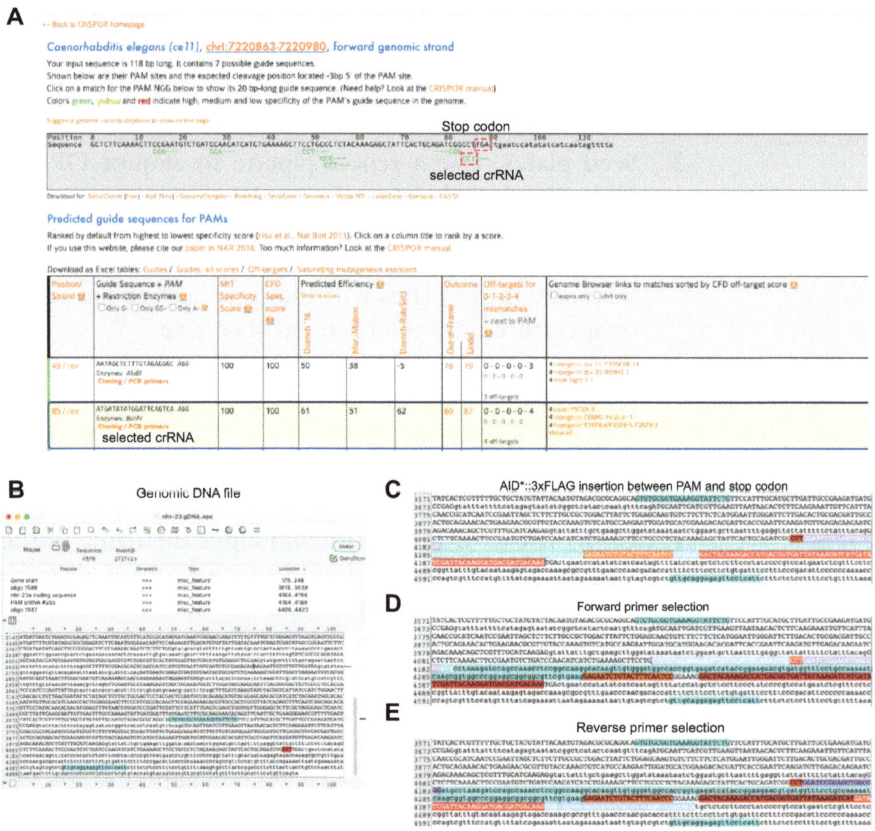

Fig. 1 crRNA selection and knock-in design. (**A**) crRNA selection via CRISPOR for C-terminal tagging of NHR-23. The selected crRNA and position of the stop codon are indicated. The dashed lines following the protospacer adjacent motif (PAM) indicate the position of the DNA double-strand break produced by editing using the indicated crRNA. In this case, a crRNA immediately adjacent to the stop codon was selected. (**B**) Example of a genomic DNA file in APE. The genotyping oligos are highlighted in blue, and crRNA PAM is in red. Each has a unique numerical indicator for tracking in our oligo and crRNA databases. (**C**) Insertion of an AID*::3xFLAG cassette (highlighted) from pJW1354 immediately adjacent to the PAM. The insertion separates the PAM from the crRNA binding sequence, preventing re-cutting after insertion. The recommended forward oligo (**D**; highlighted) and reverse oligo (**E**, highlighted). Each has 35 nt of sequence flanking the insertion at the 5′ end of the primer. Note that for E, one would need to take the reverse complementary sequence.

on either side of the desired insertion site. Paste this sequence into APE and save it in the folder as the "gene name gDNA."

3. Copy ~50 bp of sequence flanking the edit site. Go to http://crispor.tefor.net/ and paste in this sequence. Select the *ce11* genome and *20bp-NGG Sp Cas9, Sp Cas9-HF1,* and *eSp Cas9 1.1* and click "Submit."

4. Pick high-scoring target sites (Fig. 1A). Ideally, look for 5′-(N18)-GG-NGG-3′ target sites. These have been the most reliably efficient sites in *C. elegans.* The next best are 5′-(N19)-G-NGG-3′ sites. Avoid crRNAs with a cytosine in the final

base before the PAM whenever possible, as these tend to have lower activity in *C. elegans* [22].

5. Once one chooses a crRNA site, assign it a number and name in your database and annotate the PAM in your genomic DNA file. Once a crRNA site is chosen, assign it a number and name in your database and annotate the PAM in your genomic DNA file. We maintain unique crRNA and oligo identifiers in genomic DNA files to facilitate reagent tracking.

3.6 Design Genotyping Oligos to Identify Knock-Ins

1. Copy 500–800 bp of the sequence flanking the insertion site from the gDNA sequence file and go to NCBI Primer BLAST https://www.ncbi.nlm.nih.gov/tools/primer-blast/.
2. Set the product size to 500–800 bp (PCR product size Min and Max). Change the database from *RefSeq mRNA* to *RefSeq representative genomes* and set the organism to *Caenorhabditis elegans*. Click "Get primers." A results page will appear showing the best matches for the input sequence. Select the gene of interest and submit. *See* **Note 8** for comments on some potential challenges with genotyping oligo design.
3. Select primers and then search for them in the genomic DNA sequence file to ensure that they flank the insertion.
4. Enter the primers in one's oligo database. We suggest recording product size in WT, melting temperature, and the source of the primer design.

3.7 AID*::3xFLAG Repair Template Design

1. Go to the annotated genomic DNA file. Identify the open reading frame. This is straightforward for N-terminal and C-terminal fusions (start and stop codons, respectively). For internal tagging, annotate the start or end of domains to guide orientation and reading frame.
2. Copy the AID*::3xFLAG sequence from pJW1354 and paste it in frame as close to the DSB as possible (Fig. 1C); Cas9 cleaves between the third and fourth base pair upstream of the PAM [23]. Similar approaches can be used to design other repair templates, such as fluorescent protein::AID*::3xFLAG knock-ins.
3. Double-check the crRNA site to ensure that the insertion inactivates it. Separating the PAM from the crRNA target is effective, as is inserting within the first 10 bp of the crRNA target sequence. In some cases, the insertion may unintentionally re-create a functional crRNA site.
4. For the forward oligo, aim for a 57–65°C melting temperature for the portion annealing to the repair template. For pJW1354, we use (Fig. 1D):
   5′-[35 nt of sequence flanking the insertion]-GGATCCGG AGGTGGCGGGATG-3′

For the reverse oligo, use a similar melting temperature. For pJW1354 we use (Fig. 1E):

5′-[35 nt of sequence flanking the insertion]-CTTGTCATCGTCATCCTTGTAATCGATATC-3′

Confirm that the reverse oligo sequence is a reverse complement; errors at this step are common.

5. Quality control. Copy the insert plus upstream sequence and paste it into a new APE file. Delete the introns and translate the sequence through to the stop codon. Run the protein sequence through BLASTP to confirm the frame is correct.

3.8 Generate Repair Template Through PCR

1. Set up twelve 50 μL PCRs in a sterile 1.5 mL Eppendorf tube using the recipe below (*see* **Note 9**).

   125 μL 5× Phusion Green HF Reaction Buffer.

   62.5 μL 2.5 mM dNTPs.

   62.5 μL 5 M betaine.

   6.25 μL DMSO.

   6.25 μL 5 μM forward primer.

   6.25 μL 5 μM reverse primer.

   25 μL Phusion polymerase.

   387.5 μL sterile ddH_2O.

2. Pipette up and down to mix and aliquot 50 μL into an 8- or 12-well PCR strip.
3. Run on a thermocycler: 95°C 3 min; 35 cycles of [95°C 30 s, 57–65°C 30 s, 72°C 45 s]; 72°C 5 min; 8°C hold.
4. Pool and run 5 μL on 1% TAE agarose gel with GelRed (or ethidium bromide) and confirm a product of the correct size.

3.9 Purify and Concentrate Repair Template

1. Resuspend AMPureXP beads by shaking. We use a 0.6× ratio of beads:PCR (i.e., 72 μL beads for 120 μL pooled PCR). Estimate pooled volume using a P200 pipette. Mix beads and PCR product thoroughly by pipetting ten times, then incubate for 5 min at room temperature.
2. Place tubes on a magnetic rack. Ensure the tube bottom is close to the magnet. Incubate for 5 min or until the solution clears.
3. Remove supernatant carefully. Wash beads twice with 500 μL of freshly diluted 70% ethanol, incubating 30 s each time and then placing the tube back on the magnet for 5 min or until the solution clears. Aspirate and discard ethanol. The beads are not drawn out easily when in alcohol, so it is not necessary to leave any supernatant behind.

4. Dry beads by incubating open tubes at 37°C for ~10 min, or until ethanol has fully evaporated.
5. Remove tubes from the magnet and elute DNA by adding 40 µL of nuclease-free water. Mix by pipetting up and down ten times, incubate for 2 min, then separate beads on a magnetic rack for 1 min or until the solution clears. Transfer eluate to a fresh tube and measure concentration with a Nanodrop.
6. Confirm there is a single PCR product. Add 3 µL of Milli-Q and 1 µL of 6× loading dye to ~2 µL of eluted PCR product and run on a 1% TAE agarose gel with GelRed (or ethidium bromide).

3.10 Microinjection Using Cas9 Ribonucleoprotein Complexes (RNPs)

Goal: Introduce the AID*::3xFLAG-tagged allele into a TIR1-expressing strain. If depletion across multiple TIR1 backgrounds is desired, generate the knock-in in wild-type N2 and cross subsequently to a strain carrying a TIR1 transgene. If only pan-somatic or tissue-specific depletion is needed, perform the knock-in directly in a strain carrying the desired TIR1 transgene. For labs that are new to *C. elegans* culture, these video-based resources are useful:
https://www.youtube.com/@WorkingwiththeWorm
https://www.youtube.com/@hitoshisawa9622

3.10.1 Preparing Animals for Injection

1. Four days before injection, pick 7 L4s to a 6 cm plate. Alternatively, pick seven adults 3 days prior to injection. Incubate at 20°C.
2. The day before injection, pick 20 L4 larvae of the desired genotype per planned injection. Inject 12–18 animals and incubate overnight at 20°C.

3.10.2 Injection Mix Preparation

1. On the day of injection, thaw an aliquot of trcRNA and the appropriate crRNA(s). Make trcRNA and Cas9 working stocks (see Sect. 2.3); these can be stored at −20°C and reused for several weeks.
2. Melt PCR repair template. Pipette 550 ng of PCR product into a PCR tube or strip. The target concentration in the mix is 50 ng/µL (500 ng total), and we include an excess to account for pipetting errors/inaccuracy. Run melting program on a thermocycler:

   95°C—2:00 (initial denaturation)

   85°C—0:10

   Ramp down at 1.0°C/s to 75°C, hold 10 s

   Ramp down at 1.0°C/s to 65°C, hold 10 s

   Ramp down at 1.0°C/s to 55°C, hold 10 s

   Ramp down at 1.0°C/s to 45°C, hold 10 s

Ramp down at 1.0°C/s to 35°C, hold 10 s

Ramp down at 1.0°C/s to 25°C, hold 10 s

Ramp down at 1.0°C/s to 4°C—hold (∞) or until you remove the plate/tubes.

3. Set up 10 μL microinjection reactions. Add 200 ng of crRNA first, as it will be the largest volume. Then add 400 ng of trcRNA and 2500 ng of Cas9. *See* **Note 10** for a sample injection mix. Pipette up and down to mix and then incubate at 37°C for 15 min in a thermocycler.
4. For each injection, one needs two 1.5 mL tubes. After incubation of the RNP and melting of the repair template, they are combined in the 1.5 mL Eppendorf tube along with any co-injection marker (*see* **Note 11**) and water to bring the reaction volume to 10 μL. The solution is mixed by pipetting up and down. The tube is then spun at 21,130 rcf for 5–10 min. The top 3–5 μL of liquid is transferred to a fresh tube to remove any precipitates or debris that could clog the needle. This solution is then used for microinjection.

3.10.3 Microinjection

1. Prepare 6-well plates; one plate per injection. Label the side of the plate with the date, strain, and injection mix. Gather supplies (worm pick, microcapillary needles, and agarose pads).
2. Microinject the parental animal in one or both gonads. Microinjection in *C. elegans* is technically demanding. We recommend video-based training protocols for new researchers [24].
3. Put plates on the bench and screen 3 days later (*see* **Note 12**).

3.10.4 PCR Screening F1 Animals for Knock-Ins (See Note 13)

1. Pick 24 F1 animals carrying the co-injection marker at the L4 stage into individual wells of a two 12-well MYOB plate. Incubate 1–2 days at 20°C until more than 25–50 eggs have been laid and then genotype the F1 parent.
2. Make lysis buffer. We use 10 μL per animal to be genotyped.

   24 reaction lysis buffer:

   25 μL 10× Phusion buffer (homemade).

   220 μL sterile Milli-Q water.

   5 μL Proteinase K (20 mg/mL).

   Pipette up and down to mix and then aliquot 10 μL per tube; we prefer 8- or 12-well PCR strips.
3. Pick the parental F1 animal into the tube. Be sure to label the tube numbers and wells on the plate so that one can trace the genotyping data back to a specific well. Freeze on dry ice for at least 10–30 min or flash freeze in liquid nitrogen.

4. Run on a lysis program in a thermocycler (65°C 90 min, 95°C 15 min). Store on ice until ready to proceed. Alternatively, lysates can be stored at −80°C indefinitely.
5. Set up a PCR master mix without a template. A 3× master mix is sufficient for 12 × 10 μL genotyping PCRs. Combine the following:

   30 μL 5× Phusion Green HF Reaction Buffer.

   15 μL 2.5 mM dNTPs.

   15 μL 5 M betaine.

   3 μL DMSO.

   3 μL 5 μM forward primer.

   3 μL 5 μM reverse primer.

   6 μL Phusion.

   287.5 μL sterile ddH_2O.

   Pipette up and down to mix and aliquot 10 μL per tube in a 12-well PCR strip. Mark the strip to match the lysate strip so that one can match a PCR reaction to a well in the MYOB plate. One can scale the reaction up or down as needed.
6. Add 1 μL of lysate to each PCR reaction. Add a cap.
7. Run on a thermocycler. Our typical reaction is:
   95°C 2 min; 35 cycles [95°C 30 s, melting temp (57–65°C usually) 30 s, 72°C (30 s/kb)]; 72°C 5 min, 12°C hold. Store reactions at 4°C until ready to run.
8. Pour a 1% agarose-TAE gel with a DNA stain. We prefer GelRed, though ethidium bromide is an alternative. We typically use 26-well combs, which are compatible with 12-channel pipettes.
9. Run reactions for 1 h at 100 V. Image on one's preferred gel imaging system. Export image and annotate, looking for a band shift of the expected insertion size. For AID*::3xFLAG amplified from pJW1354, the insertion is 246 bp (*see* **Note 14**).

3.11 Auxin Treatment of C. elegans *on Plates*

This protocol is for a small-scale pilot experiment. For larger-scale experiments, alkaline bleaching produces more animals (protocol available at: https://doi.org/10.17504/protocols.io.j8nlkkyxdl5r/v1).

1. Pick ten adult animals of the desired genotype onto a seeded well in a 6-well plate. When working with a new AID*-tagged allele, we suggest plating on both ethanol control and auxin plates. Initially, we suggest to pick:
   a. Control strain with TIR1 transgene alone
   b. Control strain with AID*-tagged allele alone
   c. Experimental strain with both TIR1 transgene and AID*-tagged allele

2. Let parents lay eggs for 30–60 min.
3. Remove parents with a platinum worm pick, sterilizing the pick in a flame. Be sure to remove all parental animals.
4. Incubate plate at 20°C and score daily for plate-level phenotypes (*see* **Note 15**).

4 Notes

1. Since we do not use LB+strep agar plates very frequently, we add 30 µL of 1000× streptomycin to 150 µL of sterile Milli-Q water and spread it on an LB plate. Allow the liquid to absorb into the plate. Since our plates are ~30 mL, this produces a 50 µg/mL LB+streptomycin agar plate.
2. We prefer MYOB, as no additional chemicals need to be added post-autoclaving. As our lab has undergraduates pour plates with frequent turnover, we find we have less contamination with MYOB compared to NGM. However, our protocol is readily ported over to NGM media of an equal volume.
3. KRY88 timed synchronized L1s should arrest with complete penetrance in late L1. They eventually attempt to molt, which fails, leading to lethality [5]. JDW92 synchronized L1s should reach adulthood. Animals will lay unfertilized eggs, which fail to hatch, and the phenotype should be completely penetrant [25]. When auxin plates start going off, both strains will produce viable progeny, with KRY88 progeny potentially also displaying a dumpy phenotype.
4. We frequently prepare MYOB in glass bottles in 250 mL or 500 mL volumes. This MYOB agar can be microwaved to melt the media; it can be cooled to 55°C on a hot plate with a stir bar, and then auxin or ethanol can be added. Desired volumes can be dispensed into sterile Petri dishes or plates with an electronic pipette and a 35 or 50 mL sterile disposable pipette.
5. 1.6 M auxin solution is at the limit of solubility. Make a solution right before one is ready to add the auxin to the MYOB at 55°C. Heat water to 55°C on a hot plate using a stir bar to distribute the heat. Add the auxin to ethanol in a 15 mL Falcon tube and put it in the water for 5 min. Vortex to mix and immediately add to the MYOB. Alternatively, one can make a 400 mM IAA stock and add 4× more to the MYOB. Note that this requires making 1% ethanol control plates. We use the more concentrated stock so that we can make control plates at a single concentration of ethanol.

6. If one does not have a PourBoy, one can use alternate plate pouring systems or an electronic pipette with a 35 or 50 mL sterile disposable pipette.
7. We only use a hydrogen peroxide solution for 3 months; after that, its effectiveness wanes, and we observe more contaminated plates. When we open a new bottle of hydrogen peroxide, we add lab tape and label when the bottle was opened and when it should be disposed of.
8. Duplicated genes can be challenging to genotype. These might require using a larger sequence range to design genotyping primers. Having one or both primers in non-coding sequence often helps improve primer specificity, as this sequence is often less conserved.
9. For repair templates that prove difficult to amplify, we suggest adding betaine and DMSO and reducing the amount of water accordingly.
   a. 5 M betaine: add 5 μL per 50 μL PCR reaction.
   b. DMSO: add 1 μL per 50 μL PCR reaction.
10. Sample injection mix

10,000 ng/μL IDT Cas9 (1:4 dilution to 2500 ng/μl in 1× PBS)	1.0 μL	250 ng/μL final
510 ng/μL New #255 crRNA (Diluted to 200 ng/μl)	1.0 μL	20 ng/μL final
3350 ng/μL trcRNA (1:10 dilution to 400 ng/μl in 1× PBS)	1.0 μL	40 ng/μL final

Incubate for 15 min at 37°C.

126 ng/μL pSEM229 *mlc-1p::RFP-NLS* (Invitrogen purified)	1.9 μL	20 ng/μL final
1270 ng/μL PCR #746 (Diluted to 500 ng/μL)	1.0 μL	50 ng/μL final
Nuclease-free water	4.1 μL	10 μL total

11. We frequently use pCFJ90 (*myo-2p::mCherry*; https://www.addgene.org/19327/) [26] at 5 ng/μL or pSEM229 (*mlc-1p::mNeonGreen*; https://www.addgene.org/159896/) [27] at 25 ng/μL.
12. The majority of edits for RNP-based methods have been shown to occur 24 h post-injection [28], so one can consider

moving the injected P0 animal to a new plate after 24 h and screening the new plate.

13. For fluorescent protein::AID cassette knock-ins into strongly expressed genes, one can often recover the animals under a fluorescent stereomicroscope.
14. If one fails to recover a knock-in, re-attempt and screen more animals. One can also take Sanger sequencing.ab1 files and run them through Synthego's ICE Tool, which searches for signatures of non-homologous end joining, which would indicate that a DNA double-strand break was formed. If ICE suggests that a break was not formed, this data suggests that the crRNA/sgRNA was ineffective, and a new crRNA targeting a different sequence should be used.
https://www.synthego.com/products/bioinformatics/analysis
15. Determining the extent and rate of depletion is desirable. If performing pan-somatic depletion, or depletion of a protein restricted to a specific tissue, then Western blotting is a useful approach. Should one use a fluorescent protein::AID* construct, then microscopy can be used. This latter approach is better suited for tissue-specific depletion of a protein expressed in multiple cell types.

Acknowledgments

I thank Dr. James Matthew Ragle and Dr. Guinevere Ashley for feedback on this manuscript. This work was supported by NIH/NIGMS awards R01GM138701 and R35GM158317.

References

1. Ruegger M, Dewey E, Gray WM, Hobbie L, Turner J, Estelle M (1998) The TIR1 protein of *Arabidopsis* functions in auxin response and is related to human SKP2 and yeast Grr1p. Genes Dev 12:198–207
2. Leyser O (2018) Auxin signaling. Plant Physiol 176:465–479. https://doi.org/10.1104/pp.17.00765
3. Dharmasiri N, Dharmasiri S, Estelle M (2005) The F-box protein TIR1 is an auxin receptor. Nature 435:441–445. https://doi.org/10.1038/nature03543
4. Tan X, Calderon-Villalobos LIA, Sharon M, Zheng C, Robinson CV, Estelle M et al (2007) Mechanism of auxin perception by the TIR1 ubiquitin ligase. Nature 446:640–645. https://doi.org/10.1038/nature05731
5. Zhang L, Ward JD, Cheng Z, Dernburg AF (2015) The auxin-inducible degradation (AID) system enables versatile conditional protein depletion in *C. elegans*. Development 142:4374–4384. https://doi.org/10.1242/dev.129635
6. Nishimura K, Fukagawa T, Takisawa H, Kakimoto T, Kanemaki M (2009) An auxin-based degron system for the rapid depletion of proteins in nonplant cells. Nat Methods 6:917–922. https://doi.org/10.1038/nmeth.1401
7. Holland AJ, Fachinetti D, Han JS, Cleveland DW (2012) Inducible, reversible system for the rapid and complete degradation of proteins in mammalian cells. Proc Natl Acad Sci U S A 109:E3350–E3357. https://doi.org/10.1073/pnas.1216880109

8. Natsume T, Kiyomitsu T, Saga Y, Kanemaki MT (2016) Rapid protein depletion in human cells by auxin-inducible degron tagging with short homology donors. Cell Rep 15:210–218. https://doi.org/10.1016/j.celrep.2016.03.001
9. Ashley GE, Duong T, Levenson MT, Martinez MAQ, Johnson LC, Hibshman JD et al (2021) An expanded auxin-inducible degron toolkit for *Caenorhabditis elegans*. Genetics 217:iyab006. https://doi.org/10.1093/genetics/iyab006
10. Trost M, Blattner AC, Leo S, Lehner CF (2016) *Drosophila dany* is essential for transcriptional control and nuclear architecture in spermatocytes. Development 143:2664–2676. https://doi.org/10.1242/dev.134759
11. Brown KM, Long S, Sibley LD (2017) Plasma membrane association by N-Acylation governs PKG function in *Toxoplasma gondii*. mBio 8:e00375–17. https://doi.org/10.1128/mBio.00375-17
12. Daniel K, Icha J, Horenburg C, Müller D, Norden C, Mansfeld J (2018) Conditional control of fluorescent protein degradation by an auxin-dependent nanobody. Nat Commun 9:3297–13.https://doi.org/10.1038/s41467-018-05855-5
13. Chen W, Werdann M, Zhang Y (2018) The auxin-inducible degradation system enables conditional PERIOD protein depletion in the nervous system of *Drosophila melanogaster*. FEBS J 285:4378–4393. https://doi.org/10.1111/febs.14677
14. Camlin NJ, Evans JP (2019) Auxin-inducible protein degradation as a novel approach for protein depletion and reverse genetic discoveries in mammalian oocytes. Biol Reprod 101:704–718. https://doi.org/10.1093/biolre/ioz113
15. Sepers JJ, Verstappen NHM, Vo AA, Ragle JM, Ruijtenberg S, Ward JD et al (2022) The mIAA7 degron improves auxin-mediated degradation in *Caenorhabditis elegans*. G3 (Bethesda) 12:jkac222. https://doi.org/10.1093/g3journal/jkac222
16. Negishi T, Asakawa M, Kanemaki MT, Sawa H (2019) Modified auxin improves the auxin-inducible degradation (AID) system for laid C. elegans embryos. MicroPubl Biol 2019:1–4. https://doi.org/10.17912/micropub.biology.000190
17. Ward JD, Reiner DJ, Daul AL, Rougvie AE (2026) The Caenorhabditis Genetics Center curated special collections: a guide to protein degradation systems. Genetics 232(4):iyaf256. https://doi.org/10.1093/genetics/iyaf256
18. Yesbolatova A, Saito Y, Kitamoto N, Makino-Itou H, Ajima R, Nakano R et al (2020) The auxin-inducible degron 2 technology provides sharp degradation control in yeast, mammalian cells, and mice. Nat Commun 11: 5701. https://doi.org/10.1038/s41467-020-19532-z
19. Hills-Muckey K, Martinez MAQ, Stec N, Hebbar S, Saldanha J, Medwig-Kinney TN et al (2022) An engineered, orthogonal auxin analog/AtTIR1(F79G) pairing improves both specificity and efficacy of the auxin degradation system in *Caenorhabditis elegans*. Genetics 220:iyab174. https://doi.org/10.1093/genetics/iyab174
20. Negishi T, Kitagawa S, Horii N, Tanaka Y, Haruta N, Sugimoto A et al. (2022) The auxin-inducible degron 2 (AID2) system enables controlled protein knockdown during embryogenesis and development in *Caenorhabditis elegans*. Genetics 220:iyab218. https://doi.org/10.1093/genetics/iyab218
21. Vicencio J, Chihara D, Eder M, Sedlackova L, Ahringer J, Stroustrup N (2025) Engineering the auxin-inducible degron system for tunable *in vivo* control of organismal physiology. Nat Commun 16: 10848. https://doi.org/10.1038/s41467-025-66347-x
22. Paix A, Wang Y, Smith HE, Lee C-YS, Calidas D, Lu T et al (2014) Scalable and versatile genome editing using linear DNAs with microhomology to Cas9 Sites in *Caenorhabditis elegans*. Genetics 198:1347–1356. https://doi.org/10.1534/genetics.114.170423
23. Jinek M, Chylinski K, Fonfara I, Hauer M, Doudna JA, Charpentier E (2012) A Programmable dual-RNA-guided DNA endonuclease in adaptive bacterial immunity. Science 337:816–821. https://doi.org/10.1126/science.1225829
24. Berkowitz LA, Knight AL, Caldwell GA, Caldwell KA (2008) Generation of stable transgenic *C. elegans* using microinjection. J Vis Exp 18:e833. https://doi.org/10.3791/833
25. Ragle JM, Aita AL, Morrison KN, Martinez-Mendez R, Saeger HN, Ashley GA et al (2020) The conserved molting/circadian rhythm regulator NHR-23/NR1F1 serves as an essential co-regulator of *C. elegans* spermatogenesis. Development 147:dev193862. https://doi.org/10.1242/dev.193862
26. Frøkjaer-Jensen C, Davis MW, Hopkins CE, Newman BJ, Thummel JM, Olesen S-P et al (2008) Single-copy insertion of transgenes in *Caenorhabditis elegans*. Nat Genet 40:1375–1383. https://doi.org/10.1038/ng.248
27. El Mouridi S, Peng Y, Frøkjær-Jensen C (2020) Characterizing a strong pan-muscular promoter (Pmlc-1) as a fluorescent co-injection marker to select for single-copy insertions. MicroPubl Biol. https://doi.org/10.17912/micropub.biology.000302 PMID: 32908967; PMCID: PMC7474950
28. Ghanta KS, Mello CC (2020) Melting dsDNA donor molecules greatly improves precision genome editing in *Caenorhabditis elegans*. Genetics 216:643–650. https://doi.org/10.1534/genetics.120.303564

Chapter 13

An Auxin-Inducible Degron System for Trypanosomes

Alexandra Klein, Alice McDowell, Christian R. S. Reis, Rodrigo Pontes de Lima, Danielle M. N. Moura, Janaina de Freitas Nascimento, Osvaldo P. de Melo Neto and Mark Carrington

Abstract

The ability to specifically degrade one protein in a cell provides an immediate insight into any resulting phenotype and thus function that may be occluded by secondary effects using other techniques. The auxin-inducible degron is a method developed to achieve this end and is effective in mammalian and yeast cells. The approach is dependent on the recognition of a tagged target protein by an endogenous E3 ubiquitin ligase modified by the expression of a plant F-box protein. This approach was initially unsuccessful in trypanosomes, probably due to a lack of interaction between the endogenous E3 ubiquitin ligase and the plant F-box protein. This was overcome by expressing the entire rice E3 ubiquitin ligase in trypanosomes, and here we describe the production of the auxin-inducible degron competent cell lines and tagging of target genes.

Keywords Trypanosome, Auxin-inducible degron

1 Introduction

The ability to specifically ablate a protein still underlies the majority of investigations into the function of a particular gene. The first approach used to identify a role for a gene in a process involved making null or conditional mutants through random mutagenesis and screening for a particular phenotype. This type of forward genetics, in various guises, remains a powerful tool as it requires no knowledge of the identity of the gene. The advent of widely available genome sequences allowed the identification of target genes that may or may not be involved in a process of interest and led to an explosion in reverse genetics. In this approach, the gene was identified from sequence, altered in some way, and the resulting phenotype investigated. Any phenotype usually results from the loss of a protein in a cell, and this can be achieved at three levels: deletion of the gene, ablation of the mRNA by RNAi, or targeted degradation of

Maria W. Górna et al. (eds.), *Degrons*, Methods in Molecular Biology, vol. 3069,
https://doi.org/10.1007/978-1-0716-5508-5_13,

the protein. All are effective but each has disadvantages; recovery after gene deletion may result in a compensatory change that masks a phenotype. RNAi-based degradation of a specific mRNA occurs rapidly but not always completely; also, the protein lingers on and often has to drop below a threshold before a phenotype becomes apparent. This makes it difficult to separate primary and secondary phenotypes. Protein ablation through targeted degradation is not always complete but is rapid and provides an opportunity to identify a primary phenotype.

The African trypanosome *Trypanosoma brucei* is the causal agent of disease in humans, livestock, and wild animals. It has been intensively studied for more than 100 years, and research now centers on both the causes of pathology and basic cellular and molecular processes of a divergent eukaryote. Forward genetics is nontrivial, as trypanosomes are diploid, crosses occur in the salivary glands of the tsetse fly, and are a non-obligatory part of the life cycle. The genome was sequenced 20 years ago [1], and this opened up the use of reverse genetics as an approach to address questions about trypanosome biology. The dominant approach has been RNAi [2], which has been successful both in identifying the role of individual genes and has been modified for forward genetics using whole-genome RNAi libraries [3, 4]. The RNAi approach usually uses tetracycline-inducible promoters to produce double-stranded RNA, and the target mRNA is dramatically reduced within 1 h [2]. However, in many experiments, a phenotype does not appear for another 24 h or more (e.g., [5]), and this has been interpreted as the time taken for the protein of interest to drop below a threshold by dilution resulting from cell proliferation. This lag means it can be difficult to separate primary and secondary phenotypes.

The time taken for protein levels to diminish is a particular problem in experiments investigating mRNA turnover, as the half-life of mRNAs in trypanosomes ranges from a few minutes to a couple of hours [6]. The time taken for protein depletion after RNAi induction means it is difficult to resolve which mRNAs are altered as a primary effect of the RNAi knock down and which changes are secondary. To get around this problem, we have adapted the auxin degron system for use in trypanosomes, and here we describe its use in procyclic form *T. brucei*.

The auxin-dependent degron is based on an Skp1-Cullin-F-box (SCF) E3 ubiquitin ligase that contains TIR1 as the F-box protein [7]. TIR1 determines specificity by binding to a substrate protein in an auxin (indole-3-acetic acid, IAA)-dependent manner. The degron system was developed for use in yeast and mammalian cells and involved two steps. First, ectopic expression of TIR1 and second, the tagging of a target protein with an auxin-induced degradation (AID) domain from the plant IAA17/AXR3-1 protein that is sufficient for binding to TIR1 in an auxin-dependent manner [7]. On the addition of exogenous auxin, the target protein is ubiquitinylated

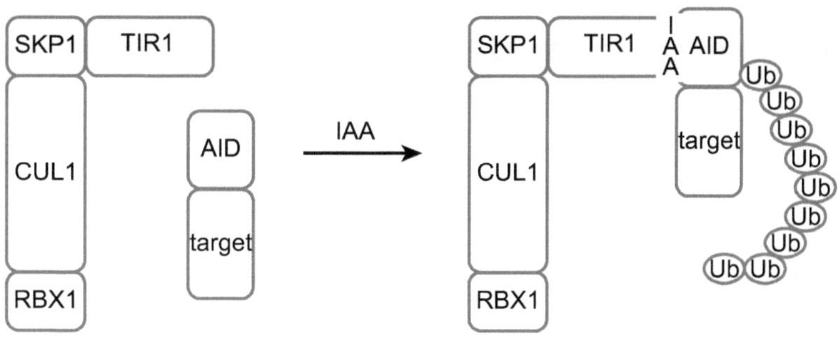

Fig. 1 Schematic illustration of the structure of the rice auxin-dependent E3 ubiquitin ligase and the binding of a protein tagged with an AID domain on the addition of the auxin indole-3-acetic acid (IAA)

and subsequently degraded by the proteosome (Fig. 1). There is no known equivalent of the auxin-dependent SCF in fungi or mammals, and for the system to work, the plant TIR1 has to bind the SKP1 component of the endogenous SCF. In addition, the endogenous levels of IAA in the experimental system have to be below the threshold for activating the SCF-TIR1.

Initial experiments with procyclic-form *T. brucei* in culture were unsuccessful, most likely due to a failure of TIR1 to interact with the endogenous SKP1. After a lot of faffing about, we decided to reconstruct the entire plant SCF-TIR1 in trypanosomes using the sequences of the components from rice, as trypanosomes are cultured at 28°C or 37°C depending on the developmental stage. However, cell lines expressing these transgenes were only partially effective. Tagging of target proteins with AID domains in cell lines expressing the rice E3 ubiquitin ligase resulted in decreased expression levels in the absence of exogenous auxin. However, there was a glimmer of hope, as addition of IAA resulted in degradation of the residual protein. The reduced level of the tagged protein is almost certainly due to levels of endogenous auxin produced by the metabolism of tryptophan [8]. Success followed the removal of this problem by altering the IAA-binding site on TIR1 through changing phenylalanine 74 to glycine and using 5-phenyl-IAA as the inducer [9] (*see* **Note 1**).

The above serves as background to the methods below, which describe first how to make a trypanosome cell line expressing rice SCF-TIR1 plus ARF, and second how to then produce a cell line with a specific gene tagged with a degron. The protocols are based on *T. brucei* Lister 427 procyclic forms but are applicable to other cell lines and developmental stages (*see* **Note 2**).

The open reading frames of four rice proteins, CUL1 (cullin/CDC53), SKP1, RBX1, and TIR1, were modified to include a single Ty epitope tag at the C-termini and were made synthetically after being codon-optimized to increase expression in trypanosomes. In addition, a domain from rice auxin response factor 16 (ARF16) with an N-terminal Ty epitope tag was added to the mix, as this had been reported to improve the functioning of the system [10]. The five transgenes were

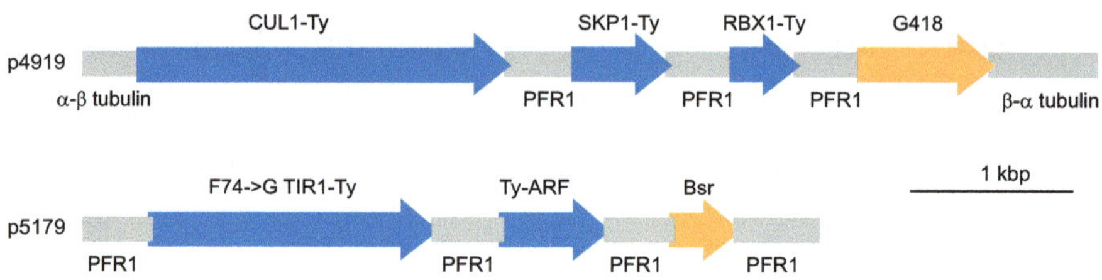

Fig. 2 Schematic illustration showing the location of the genes encoding the E3 ubiquitin ligase components and a fragment of ARF16 in the plasmids p4919 and p5179. The fragments shown are released from the plasmids after PacI digestion. PFR1 paraflagellar rod protein 1

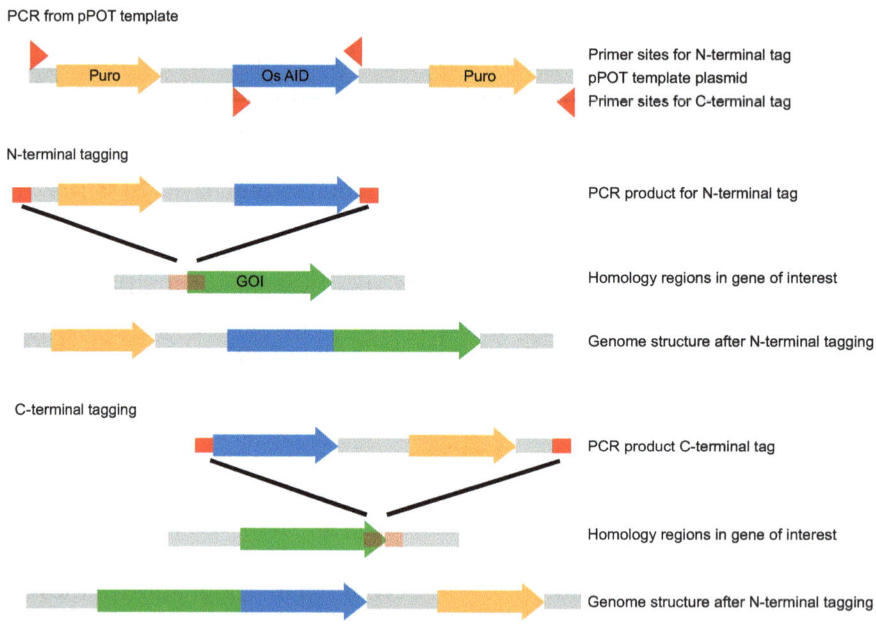

Fig. 3 Schematic illustration showing the principle behind the use of pPOT plamids [11]. PCR products contain the tag, a selectable marker cassette plus 80 bp of targeting sequence at each end. The targeting sequences are in the 100-mer primers used in the PCR

cloned into plasmid vectors in two sets: p4919 contains CUL1, SKP1, RBX1, and a G418 resistance marker; p5179 contains TIR1 F74G, ARF16 fragment, and a blasticidin resistance marker (Fig. 2). Both plasmids contain PacI sites flanking the inserts, and both were designed for read-through RNA polymerase II transcription after integration into either the tubulin locus for p4919 or the PFR1 locus for p5179.

To make a protein susceptible to auxin-inducible degradation, the gene is altered so that the encoded protein is a fusion with an AID domain. The method below is based on using derivatives of the pPOT plasmid series [11] as a template for making a PCR product that is used to modify the endogenous locus of the gene of interest (Fig. 3). Trypanosomes are diploid, and so this process is performed twice so that both alleles are modified; one set of pPOT template plasmids encodes puromycin resistance, the second hygromycin. The success of the modification can be determined by Western blotting either using an antibody recognizing the protein of interest or epitope tags added as part of the modification of the endogenous gene and/or whole-cell mass spectrometry (*see* **Note 3**).

2 Materials

Follow local safety regulations when handling *Trypanosoma brucei*.

Prepare all solutions with ultrapure water and filter-sterilize all solutions and store at 4°C.

2.1 Growth of Procyclic-Form *T. brucei* in Culture

1. *T. brucei* cell lines (ATCC).
2. SDM79 medium plus 10% heat-inactivated fetal bovine serum.
3. Incubator for growth conditions: 27°C, humidified, and 5% CO_2.
4. Range of tissue-culture flasks with vented caps.
5. 24-well plates for tissue culture.
6. Antibiotics for selection of transgenic cell lines
   a) 100× blasticidin: 1 mg/ml blasticidin.
   b) 100× G418: 1.5 mg/ml G418 (geneticin).
   c) 100× hygromycin: 2.5 mg/ml hygromycin.
   d) 100× puromycin: 0.1 mg/ml puromycin.

2.2 Preparation of Plasmids Prior to Electroporation

1. Plasmids p4919 and p5179 (Addgene 239357 and 239358).
2. PacI restriction enzyme and digestion buffer.
3. PCR Purification Kit.
4. 3 M sodium acetate buffer pH 5.0.
5. Molecular biology-grade ethanol.
6. Material and equipment for agarose gel electrophoresis of DNA.

2.3 Preparation of PCR Products Prior to Electroporation

1. Template plasmids (Addgene): pPOTv7-puro-OsAID (239359), pPOTv7-hyg-OsAID (239360), pPOTv6-puro-3Ty: OsAID:3Ty (239361), pPOTv6-hyg-3Ty:OsAID:3Ty (239362), pPOTv7-hyg-AtAID:HA (239363).
2. 100 nM 100-mer oligonucleotides.
3. Standard PCR Thermocycler.
4. Roche Expand PCR system.
5. dNTP stock solution: 10 mM each dATP, dCTP, dGTP, and dTTP.
6. Material and equipment for agarose gel electrophoresis of DNA.
7. PCR Purification Kit.
8. 3 M sodium acetate buffer pH 5.0.
9. Molecular biology-grade ethanol.

2.4 Electroporation to Make Transgenic Cell Lines

1. 15 and 50 ml sterile, conical, and capped centrifuge tubes.
2. 20 ml syringes (sterile) and 0.22 μm syringe-end filters (sterile).
3. Nucleofector and 2 mm cuvettes.
4. Sterile molecular biology-grade water.
5. 3×R buffer: 200 mM Na_2HPO_4, 70 mM NaH_2PO_4, 15 mM KCl, 150 mM HEPES pH 7.3.
6. Calcium chloride stock: 1.5 mM $CaCl_2$.
7. 50 mM 5-phenylindolylacetic acid in DMSO.
8. Materials and equipment for Western blotting.
9. Monoclonal antibodies for the detection of epitope tags: murine anti-Ty monoclonal antibody and rat anti-HA monoclonal, both commercially available.

3 Methods

3.1 Growth of Procyclic-Form T. brucei

1. Dilute the procyclic cells to 5×10^5/ml with SDM-79 medium and return to the incubator.
2. Count the cell density after 24 and 48 h. There should be a ~5-fold increase in cell density per 24 h.
3. After 48 h, dilute the cells to 5×10^5/ml and repeat incubation and counting.
4. Once the cells are proliferating at a constant rate, proceed to the electroporation. On the day before you plan to electroporate, set up a 30 ml culture at 1×10^6/ml for each electroporation.

3.2 Preparing Plasmids for Electroporation

1. The day before the electroporation, prepare a PacI digest of p4919 or p5179. Digest 10 μg of a plasmid with 50 U PacI in a 100 μl reaction following the manufacturer's instructions. Incubate at 37°C for 2 h and remove 3 μl for analysis by agarose gel electrophoresis; place the remainder of the digest at −20°C.
2. Analyze the digest using standard agarose gel electrophoresis for DNA. Both digests will contain two fragments: p4919: 2971 and 6236 bp; p5179: 2971 and 4501 bp.
3. The remaining 97 μl of the digested plasmid is then cleaned using a PCR purification kit and following the manufacturer's instructions. Elute in 200 μl into a 1.5 ml tube, then add 20 μl of 3 M sodium acetate and 550 μl of ethanol, mix by shaking. Leave at −20°C overnight.

3.3 Electroporation and Selection of Cell Lines Expressing CUL1, SKP1, and RBX1

1. Recover the digested p4919 by centrifugation in a microfuge at 15,000 × *g* for 30 min; there should be a visible pellet.
2. It is now important to keep the digested plasmid sterile, so work in a sterile safety cabinet from here on.
3. Pour away the supernatant, close the tube, and centrifuge again for 30 s; remove the residual supernatant with a 100 μl micropipette.
4. Dissolve the plasmid in 60 μl of sterile water, add 35 μl 3×R buffer and 10 μl calcium chloride stock, mix by pipetting. Keep the dissolved plasmid in the sterile safety cabinet.
5. For each electroporation, prepare three T-75 culture flasks: flask 1 containing 26 ml fresh SDM79, and flasks 2 and 3 containing 16 ml fresh SDM79.
6. Take the 30 ml of procyclic-form trypanosome culture and pipette 3 ml into a 15 ml centrifuge tube and 25 ml into a 50 ml centrifuge tube. Both tubes must have screw tops to keep the contents sterile.
7. Put the smaller tube aside for the moment.
8. Pellet the cells in the 50 ml tube at 1500 × *g* for 5 min in a benchtop centrifuge, return the tube to the sterile safety cabinet, and recover just over 20 ml of the supernatant using a 20 ml syringe. Then add a 0.22 mm filter to the syringe and add 10 ml filtered conditioned medium to flasks 2 and 3.
9. Without delay, pellet the cells in the 15 ml centrifuge tube as above. Pour away the supernatant and centrifuge again for 30 s. Remove the residual supernatant with a 1 ml pipette.
10. Resuspend the cells in 100 μl of the p4919 solution, transfer to a 2 mm cuvette, and electroporate using program X-001. Immediately transfer the electroporated cells to flask 1 and swirl to disperse.
11. Transfer 1 ml from flask 1 to flask 2, swirl, and then transfer 1 ml from flask 2 to flask 3. Place all three flasks in the growth incubator.
12. After 6 h, add 0.5 ml of 100× G418 stock to each flask (final 2× G418), swirl to mix, and then pipette 1 ml aliquots into the wells of 24-well plates. The result is three 24-well plates with dilutions 1/1, 1/25, and 1/625.
13. Return to the incubator and resist the temptation to examine the plates for 7 days.
14. After 7–10 days, selection will have occurred, and it is likely that there will be some wells in the 1/625 dilution plate with proliferating cells and others with no cells. Inspect each well using an inverted microscope, being careful to maintain sterility.

15. Mark wells with proliferating cells and note the number of wells.
16. If there are no wells with cells, look at the 1/25 plate. All the wells in the 1/1 plate should contain proliferating cells; if not, it is likely there were technical problems.
17. Once the cell density in a well is $>5\times10^6$, transfer the contents of the well into a T25 flask containing 10 ml fresh SDM79 with 0.1 ml G418 solution (1× G418). You should aim to pick wells from the highest dilution plate, as these are more likely to be clonal. Pick four or more wells to analyze.
18. Passage the cells a couple more times (Sect. 3.1) and move on to validation.

3.4 Validation

1. Prepare whole-cell lysates for Western blotting at 2×10^8 cell equivalents/ml and analyze 10 or 20 μl.
2. Probe the Western blot with anti-Ty monoclonal; typical results are shown in 3A. Select a cell line in which the three Ty-tagged proteins of the expected molecular weights are expressed and proceed.

3.5 Electroporation and Selection of Cell Lines Expressing CUL1, SKP1, RBX1, TIR1, and ARF16 Fragment

1. Proceed with the electroporation as above; start with a cell line successfully transfected with p4919 and grow in SDM79 supplemented with 1× G418 stock solution.
2. Use the PacI-digested p5179 for the electroporation as in Sect. 3.3.
3. 6 h after the electroporation, add 0.5 ml blasticidin stock (2× blasticidin) and 0.25 ml G418 stock per flask, then plate the three dilutions into 24-well plates as in Sect. 3.3.
4. Be patient and once the cell density in a well is $>5\times10^6$, transfer the contents to a T25 flask containing 10 ml fresh SDM79 with 1× blasticidin and 1× G418. It is worth picking six or more wells to analyze, and you should aim to pick wells from the highest dilution plate, as these are more likely to be clonal.
5. Passage the cells a couple more times (Sect. 3.1) and move on to validation (Sect. 3.4); example results are shown in Fig. 4.
6. Select two cell lines to test for degron ability.

3.6 Synthesis of a PCR Product for Making Degron Cell Lines

1. Design of oligonucleotide primers [11]. Two 100-mer primers are needed; the 5′ 80 nucleotides of each contain the sequence of the target site in the genome and, at the 3′ end, 20 nucleotides that base pair with the pPOT template. It should be noted that the quality of such long oligonucleotides varies with supplier.
2. PCR reaction (*see* **Note 4**). The choice of template plasmid will be determined by whether you would like to add epitope tags to your AID-tagged protein to determine depletion. You will

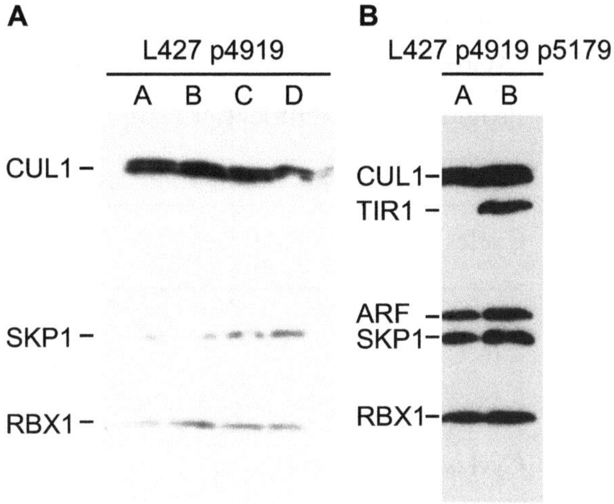

Fig. 4 Western blots of screens for the expression of E3 ubiquitin ligase components. (**A**) Four cell lines (A to D) were isolated after transfection with p4919 and selection with G418. Whole-cell lysates (1×10^6 cell equivalents) were probed with anti-Ty, and bands corresponding to the expected sizes of CUL1, SKP1, and RBX1 were observed. The strength of the signal from CUL1 is always greater for unknown reasons, and it is always a doublet, possibly due to modification with NEDD8. (**B**) Two cell lines (A and B) were isolated after transfection of cells with first p4919 and then p5179. Whole-cell lysates (4×10^6 cell equivalents) were probed with anti-Ty; cells in A did not express TIR1 and were not used further; cells in B successfully expressed all five proteins

need to assay for depletion, and in the absence of specific assay or antibodies, the use of an epitope tag works well.

3. Set up two reaction conditions for each amplification, one with 1.5 mM and the second with 2.5 mM $MgCl_2$.
   1.5 mM $MgCl_2$:

pPOT template plasmid	20 ng
Expand buffer 2	5 μl
10 mM each dNTP	2 μl
10 pmol/μl oligo 1	1.5 μl
10 pmol/μl oligo 2	1.5 μl
Water	to 49 μl

   2.5 mM $MgCl_2$:

pPOT template plasmid	20 ng
Expand buffer 3	5 μl
25 mM $MgCl_2$	5 μl
10 mM each dNTP	2 μl
10 pmol/μl oligo 1	1.5 μl

10 pmol/μl oligo 2	1.5 μl
Water	to 49 μl

4. Program the thermocycler

96°C	240 s

Cycles 1–5

94°C	15 s
46°C	60 s
68°C	60 s per kbp of product

Cycles 6–30

94°C	15 s
60°C	60 s
68°C	60 s + 5 s increase per cycle per kbp of product

After 30 cycles

68°C	420 s
4°C	indefinite

5. Add 1 μl polymerase in the first cycle by pausing the program once the temperature has decreased to 46°C.
6. At the end of the amplification reaction, analyze 2 μl of PCR product by agarose gel electrophoresis; store the remainder at −20°C.
7. You should expect a substantial amount of PCR product (Fig. 5) and aim for 5 μg or more from the 50 μl reaction. Perform additional PCR reactions if necessary.
8. Purify the PCR product using a PCR purification kit. Elute in 200 μl into a 1.5 ml tube, then add 20 μl of 3 M sodium acetate and 550 μl of ethanol, mix by shaking. Leave at −20°C overnight.

3.7 Electroporation to Tag the First Allele

1. Culture the degron-proficient cell line in SDM-79 with 1× G418 and 1× blasticidin; once it is proliferating at the expected rate, prepare 30 ml cultures at 1×10^6/ml the day before the electroporation.
2. Recover the PCR product by centrifugation in a microfuge at 15,000 × *g* for 30 min, as in Sect. 3.2. There should be a visible pellet.
3. Dissolve as for the plasmid in Sect. 3.2.

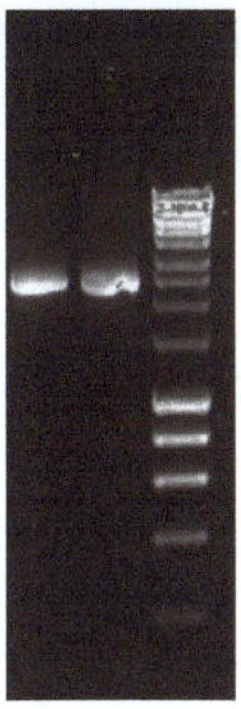

Fig. 5 Example of the PCR products used to add the AID tag by homologous recombination at the endogenous allele. The products made with 1.5 and 2.5 mM Mg^{2+} are indicated, and 2 μl of the PCR reaction were analyzed

4. For each electroporation, prepare three T-75 culture flasks: flask 1 containing 28 ml fresh SDM79, and flasks 2 and 3 containing 16 ml fresh SDM79.
5. Proceed as in Sect. 3.3 and add 10 ml conditioned medium to flasks 2 and 3.
6. Perform the electroporation as in Sect. 3.3 and add the cells to flask 1. Transfer 2.5 ml from flask 1 to flask 2, swirl to mix, then transfer 2.5 ml from flask 2 to flask 3 so you end up with 1 in 10 and 1 in 100 dilutions.
7. After 6 h, either add 0.5 ml puromycin stock (2× puromycin) or 0.5 ml hygromycin stock (2× hygromycin) to flasks 1, 2, and 3 and plate out in 24-well plates. The choice of puromycin or hygromycin is determined by the template plasmid used to make the PCR product.
8. Return to the incubator and, after 7–10 days, check each well for active proliferating cells; you should expect between 20 and 500 transformants per electroporation.
9. Once the cell density in a well is $>5 \times 10^6$, transfer the contents to a T25 flask containing 10 ml fresh SDM79 with 1× blasticidin, 1× G418, and 1× puromycin/hygromycin. It is worth picking six or more wells to analyze, and you should aim to pick wells from the highest dilution plate, as these are more likely to be clonal.
10. Passage the cells a couple more times (Sect. 3.1) and move on to validation (Sect. 3.4); example results are shown in Fig. 6A.

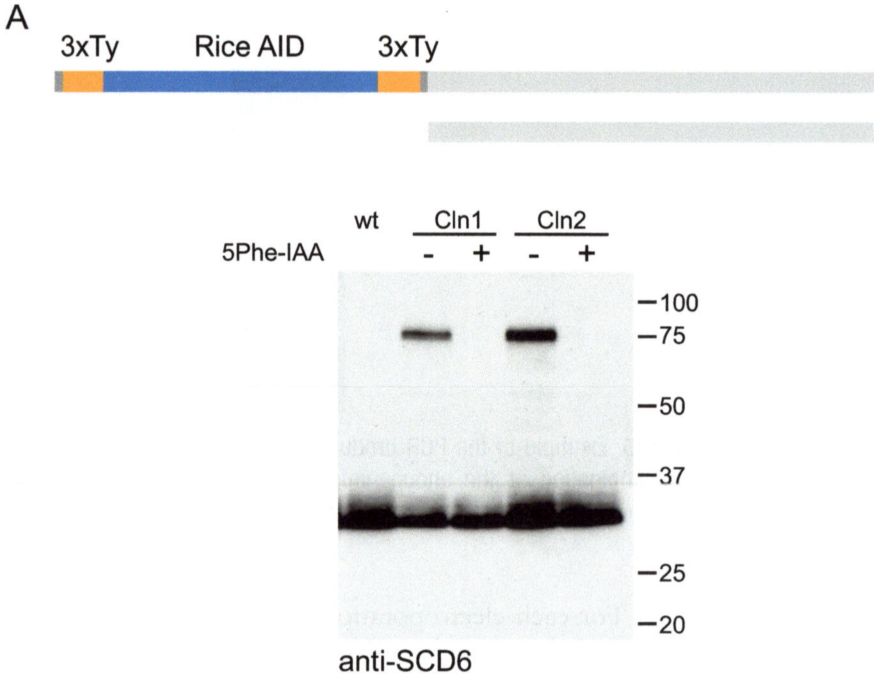

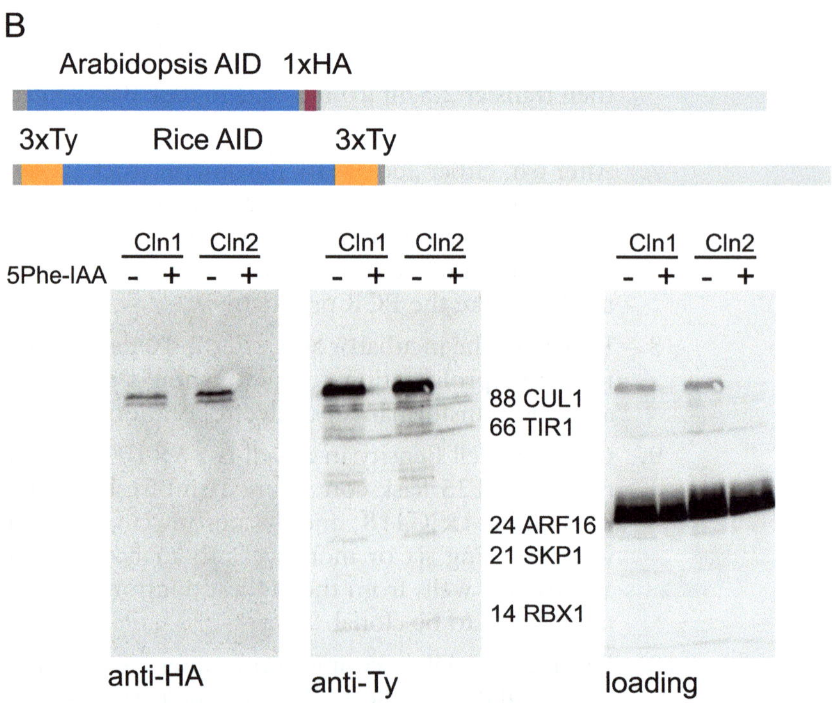

Fig. 6 (**A**) Schematic illustration showing the two alleles of the target gene and a Western blot probed with anti-SCD6 showing successful tagging of one allele of the SCD6 gene and degradation of the tagged protein 1 h after the addition of 50 μM 5Phe-IAA. 1×10^6 cell equivalents loaded per track. (**B)** Schematic showing the two

3.8 Validation of Degron Cell Lines

1. Prepare whole-cell lysates for Western blotting at 2×10^8 cell equivalents/ml and analyze 10 or 20 µl. Make lysates before and 1 h after the addition of 5Phe-IAA to 50 µM.
2. Probe the Western blot with either a protein-specific antibody or anti-Ty /anti-HA monoclonal antibody if you have added an epitope tag.
3. With a protein-specific antibody, you should observe a band corresponding to the untagged protein and one for the tagged protein. The tagged protein band should be diminished on addition of the inducer (Fig. 6A).
4. In the case of using a monoclonal antibody against the tag, you should see a single band that goes after addition of inducer. The use of a Ty-AID for the tagging of the first allele has an advantage, as it confirms the E3 ubiquitin ligase components are still being expressed (Fig. 6B).
5. Select a couple of cell lines in which the tagged protein is degraded after addition of inducer.

3.9 Electroporation of Tag the Second Allele

1. Grow the validated cells that were validated using the protocol in Sect. 3.8 with antibiotic selection; 1× G418, 1× blasticidin, and 1× puromycin or hygromycin.
2. Repeat the process described in Sect. 3.7 and select four to six wells from the most dilute plate. These should now proliferate happily in the presence of all four antibiotics.

3.10 Validation of Degron Cell Lines

1. Again, grow the cell lines with all four selection antibiotics and check for degradation of the target protein after addition of inducer by Western blotting as in Sect. 3.8.
2. If a protein-specific antibody is available, then the Western blot will show if both alleles are tagged and whether the protein is degraded on addition of inducer.
3. In the absence of a specific antibody, probe the blots sequentially with anti-HA, then anti-Ty, and then a third antibody as a loading control (Fig. 6B). This provides evidence that both alleles are tagged and degraded.

3.11 Validation by Mass Spectroscopy

A second approach is to use whole-cell mass spectrometry before and after addition of 5Phe-IAA, although this may not be sensitive enough with low-copy-number proteins.

Fig. 6 (continued) alleles of target gene and Western blots illustrating the use of epitope tags to demonstrate tagging of both alleles and degradation of the targeted protein. 2×10^6 cell equivalents were loaded per track. One allele was modified with an *Arabidopsis* AID domain containing a single HA epitope; the second was modified with a rice AID domain containing six Ty epitopes. Both are degraded after the addition of 5Phe-IAA

4 Notes

1. The rice E3 ubiquitin ligase is present in the cytoplasm and so will only work with those accessible proteins, probably just the cytoplasm and nucleus. It will not work for proteins inside the endosome. At this stage, success with proteins in transit to the mitochondrion or transmembrane proteins with a small cytoplasmic domain remains unknown. We have used the degron cell line with around 15 different cytoplasmic proteins, and all have been successfully degraded on induction.
2. The approach is dependent on the AID tag being tolerated by the cell.
3. The system should be easily transferable to other trypanosomatid species.
4. We have used the Roche Expand system for these reactions, but other proofreading thermostable polymerases should work as well.

Acknowledgments

MC is a Wellcome Investigator (217138/Z/19/Z). RPdL, CR, and DNM were funded by a CNPq postdoctoral international fellowship in the Science Without Borders Program. JdFN was funded by CAPES and Cambridge Overseas Trust. AM was funded by a Wellcome summer studentship. AK was funded by St John's College.

References

1. Berriman M, Ghedin E, Hertz-Fowler C, Blandin G, Renauld H, Bartholomeu DC, Lennard NJ, Caler E, Hamlin NE, Haas B, Bohme U, Hannick L, Aslett MA, Shallom J, Marcello L, Hou L, Wickstead B, Alsmark UC, Arrowsmith C, Atkin RJ, Barron AJ, Bringaud F, Brooks K, Carrington M, Cherevach I, Chillingworth TJ, Churcher C, Clark LN, Corton CH, Cronin A, Davies RM, Doggett J, Djikeng A, Feldblyum T, Field MC, Fraser A, Goodhead I, Hance Z, Harper D, Harris BR, Hauser H, Hostetler J, Ivens A, Jagels K, Johnson D, Johnson J, Jones K, Kerhornou AX, Koo H, Larke N, Landfear S, Larkin C, Leech V, Line A, Lord A, Macleod A, Mooney PJ, Moule S, Martin DM, Morgan GW, Mungall K, Norbertczak H, Ormond D, Pai G, Peacock CS, Peterson J, Quail MA, Rabbinowitsch E, Rajandream MA, Reitter C, Salzberg SL, Sanders M, Schobel S, Sharp S, Simmonds M, Simpson AJ, Tallon L, Turner CM, Tait A, Tivey AR, Van Aken S, Walker D, Wanless D, Wang S, White B, White O, Whitehead S, Woodward J, Wortman J, Adams MD, Embley TM, Gull K, Ullu E, Barry JD, Fairlamb AH, Opperdoes F, Barrell BG, Donelson JE, Hall N, Fraser CM, Melville SE, El-Sayed NM (2005) The genome of the African trypanosome Trypanosoma brucei. Science 309(5733):416–422. https://doi.org/10.1126/science.1112642
2. Ngo H, Tschudi C, Gull K, Ullu E (1998) Double-stranded RNA induces mRNA degradation in Trypanosoma brucei. Proc Natl Acad Sci U S A 95(25):14687–14692. https://doi.org/10.1073/pnas.95.25.14687
3. Alsford S, Eckert S, Baker N, Glover L, Sanchez-Flores A, Leung KF, Turner DJ, Field MC, Berriman M, Horn D (2012) High-throughput decoding of antitrypanosomal drug

efficacy and resistance. Nature 482(7384):232–236. https://doi.org/10.1038/nature10771
4. Mony BM, MacGregor P, Ivens A, Rojas F, Cowton A, Young J, Horn D, Matthews K (2014) Genome-wide dissection of the quorum sensing signalling pathway in Trypanosoma brucei. Nature 505(7485):681–685. https://doi.org/10.1038/nature12864
5. Kramer S, Marnef A, Standart N, Carrington M (2012) Inhibition of mRNA maturation in trypanosomes causes the formation of novel foci at the nuclear periphery containing cytoplasmic regulators of mRNA fate. J Cell Sci 125(Pt 12):2896–2909. https://doi.org/10.1242/jcs.099275
6. Manful T, Fadda A, Clayton C (2011) The role of the 5′-3′ exoribonuclease XRNA in transcriptome-wide mRNA degradation. RNA 17(11):2039–2047. https://doi.org/10.1261/rna.2837311
7. Nishimura K, Fukagawa T, Takisawa H, Kakimoto T, Kanemaki M (2009) An auxin-based degron system for the rapid depletion of proteins in nonplant cells. Nat Methods 6(12):917–922. https://doi.org/10.1038/nmeth.1401
8. Hall JE, Seed JR (1984) Increased urinary excretion of aromatic amino acid catabolites by Microtus montanus chronically infected with Trypanosoma brucei gambiense. Comp Biochem Physiol B 77(4):755–760. https://doi.org/10.1016/0305-0491(84)90309-2
9. Yesbolatova A, Saito Y, Kitamoto N, Makino-Itou H, Ajima R, Nakano R, Nakaoka H, Fukui K, Gamo K, Tominari Y, Takeuchi H, Saga Y, Hayashi KI, Kanemaki MT (2020) The auxin-inducible degron 2 technology provides sharp degradation control in yeast, mammalian cells, and mice. Nat Commun 11(1):5701. https://doi.org/10.1038/s41467-020-19532-z
10. Sathyan KM, McKenna BD, Anderson WD, Duarte FM, Core L, Guertin MJ (2019) An improved auxin-inducible degron system preserves native protein levels and enables rapid and specific protein depletion. Genes Dev 33(19-20):1441–1455. https://doi.org/10.1101/gad.328237.119
11. Paterou A, Saez Conde J, Tyc J, Sunter JD, Vaughan S, Gull K, Dean S (2025) A comprehensive toolkit for protein localization and functional analysis in trypanosomatids. Open Biol 15(4):240361. https://doi.org/10.1098/rsob.240361

Chapter 14

Degrade Green Fluorescent Protein (deGradFP) Method in *Trypanosoma brucei*

Midori Ishii and Bungo Akiyoshi

Abstract

Degrade green fluorescent protein (deGradFP) is a degron method that was originally developed in *Drosophila melanogaster*. It targets GFP-fusion proteins to an endogenous ubiquitin-proteasome pathway using an anti-GFP nanobody. We have adapted it to a kinetoplastid parasite *Trypanosoma brucei* to rapidly degrade kinetochore proteins that have low turnover rates. Here, we describe the workflow of deGradFP in *T. brucei*.

Keywords *Trypanosoma brucei*, Kinetoplastid, Kinetochore, Degron, deGradFP

1 Introduction

Conditional knockdown is an invaluable method for testing protein function in cells. In *Trypanosoma brucei*, the causative agent of African sleeping sickness, this has been achieved by conditional knockout using the Cre-LoxP system [1] or inducible CRISPR [2] at the gene level, and by the Tet-off system [3] or RNAi [4, 5] at the transcript level. These methods target either DNA or RNA. The efficiency of protein depletion therefore depends on the turnover of the RNA and protein, meaning that these methods cannot efficiently deplete proteins with low turnover rates. To overcome this problem, two targeted protein degradation methods were recently established in *T. brucei*, namely degrade green fluorescent protein (deGradFP) [6] and auxin-inducible degron (AID) [7] (described elsewhere in this book chapter; *see* **Note 1**).

In the deGradFP system, an N-terminal part of an F-box protein FBP75 (Tb927.5.700) fused with an anti-GFP nanobody VhhGFP4 (deGradFP construct) is ectopically expressed by a doxycycline-inducible system. GFP-fusion proteins are recognized by the anti-GFP nanobody (*see* **Note 2**), get ubiquitinated by the endogenous SCF complex, and are subsequently degraded by proteasomes (Fig. 1).

Maria W. Górna et al. (eds.), *Degrons*, Methods in Molecular Biology, vol. 3069,
https://doi.org/10.1007/978-1-0716-5508-5_14,

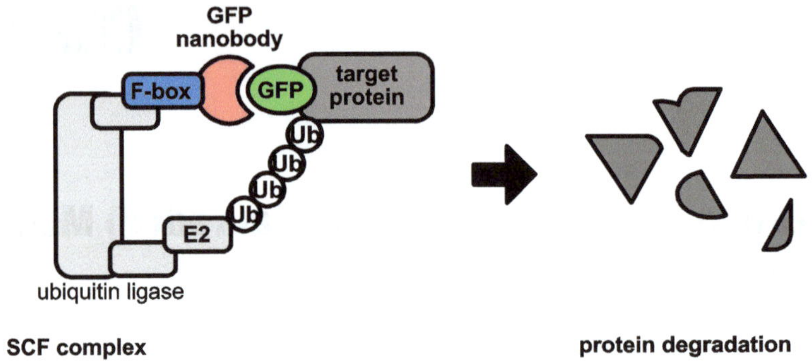

Fig. 1 Schematic illustration of the deGradFP system. (Adapted from [6] under a CC-BY 4.0 license)

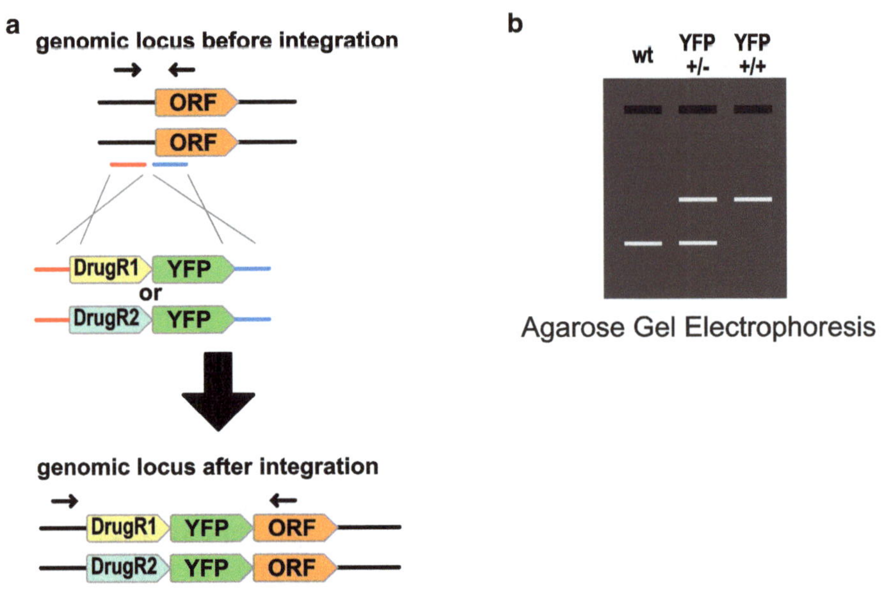

Fig. 2 (**a**) Example of N-terminal YFP-tagging and genotyping primer design. Red and blue lines show target sequences used for homologous recombination. Black arrows show the position of primers used for genotyping. (**b**) Expected bands for genotyping PCR for wild type (wt), heterozygously YFP-tagged cells (+/−), and homozygously YFP-tagged cells (+/+)

Because *T. brucei* is diploid, both alleles of a gene of interest must be tagged with GFP to study its knockdown phenotype (Fig. 2a).

2 Materials

Follow the safety rules when performing the experiments. Use ultrapure water to prepare all the solutions. For PCR and restriction enzyme reaction, use autoclaved ultrapure water.

2.1 Procyclic Form Cell Culture

1. Cell lines (*see* **Note 3**): *Trypanosoma brucei* procyclic cells with doxycycline-inducible expression system. We use either SmOx927 that expresses T7 RNA polymerase and tetracycline repressor [8] or PCF1339 (TREU927 with pJ1339) that expresses Cas9, T7 RNA polymerase, and tetracycline repressor for CRISPR-based tagging [9].
2. SDM79 media supplemented with 10% FCS. Filter-sterilized, store at 4°C.
3. Puromycin (×1000): 1 mg/mL in water, filter sterilized, store at −20°C.
4. G418 (×1000): 30 mg/mL in water, filter-sterilized, store at −20°C.
5. Hygromycin (×1000): 45–60 mg/mL solution, filter-sterilized, store at −20°C.
6. Phleomycin (×1000): 5 mg/mL in water, filter-sterilized, store at −20°C.
7. Doxycycline (×1000): 1 mg/mL in water, filter-sterilized, store at −20°C.
8. 25 cm^2 cell culture flask.
9. 96-well plate, flat-bottom.
10. 24-well plate.
11. 55 ml reagent reservoir.
12. 28°C incubator.

2.2 Transfection and Clone Isolation

1. YFP tagging system (*see* **Note 3**): For PCR-based YFP tagging, pPOTv7-eYFP series with hygromycin and G418 selection markers [10]. For plasmid-based YFP tagging, pEnT5-Y with a hygromycin selection marker and its derivative with a G418 selection marker [11].
2. pBA2675 (Inducible expression of $FBP75^{1-200}$-NLS-VhhGFP4, Addgene #189997) for nuclear proteins or pBA2705 (Inducible expression of $FBP75^{1-200}$-VhhGFP4, Addgene #189998) for cytoplasmic proteins.
3. Zimmerman post-fusion medium (ZPFM): 132 mM NaCl, 8 mM KCl, 8 mM Na_2HPO_4, 1.5 mM KH_2PO_4, 0.5 mM $Mg(CH_3COO)_2$, 0.09 mM $Ca(CH_3COO)_2$ [12]. Filter-sterilized. Store at room temperature.
4. High-fidelity DNA polymerase, such as PrimeSTAR MAX or KOD One.
5. Bio-Rad Gene Pulser Xcell electroporation systems.
6. Bio-Rad 4 mm gap Gene Pulser electroporation cuvettes.
7. Thermal cyclers and tubes for PCR.

8. 37°C incubator.
9. Molecular biology-grade ethanol.
10. *Not*I and its buffer.
11. Agarose gel electrophoresis materials and equipment.

3 Methods

3.1 Preparation of Donor DNA (Option 1): PCR-Based Tagging

1. Mix 25 μL of 2× PCR master mix (high-fidelity polymerase such as PrimeSTAR MAX or KOD One), 1 μL of 30 ng/μL of template pPOT plasmid, 1.5 μL of 10 μM forward primer, 1.5 μL of 10 μM reverse primer, 21 μL of water in a 0.2 mL PCR tube.
2. Set thermal cycler: Step 1: 98°C for 30 s. Step 2: 98°C for 10 s. Step 3: 55°C for 15 s. Step 4: 72°C for 1 min. Step 5: 30 cycles of steps 2–4. Step 6: 72°C for 30 s. Step 7: keep at 16°C.
3. Option: Run 2 μL of PCR product on 1% agarose gel to confirm PCR reaction (see Sect. 3.7). Use PCR product for transfection without any purification.

3.2 Preparation of Donor DNA (Option 2): Plasmid-Based Tagging (pEnT5-Y and Its Derivative) and deGradFP Plasmid Transfection

1. Mix 5 μg of plasmid DNA, 7 μL of 10× restriction enzyme buffer, 1 μL of *Not*I, and water to make 70 μL in a 1.5 mL tube.
2. Incubate at 37°C for 1 h.
3. Add 140 μL of 100% ethanol, mix by pipetting or inverting several times, and centrifuge at 20,000 × *g* for 10 min at room temperature (*see* **Note 4**).
4. Discard the supernatant.
5. Add 300 μL of 70% ethanol to rinse, centrifuge at 20,000 × *g* for 5 min at room temperature.
6. Discard the supernatant and dry up in a microbiological safety cabinet.
7. Resuspend DNA in 20 μL of ZPFM.

3.3 Transfection by Electroporation (Population Culture)

Transfection of the two YFP-tagging constructs can be done in two steps, i.e., first make a heterozygously YFP-tagged strain in either population or clones, then transfect another donor DNA with a different selectable marker gene and isolate homozygously tagged clones (*see* **Note 5**).

1. In the afternoon, 1 day before transfection, prepare a 10 mL cell culture (per transfection) of 2×10^6 cells/mL with selection drug(s).

2. Start transfection in the afternoon. Grow procyclic cells to 0.8–1 × 10^7 cells/mL. Use 5 mL (~5 × 10^7cells) per transfection.
3. Spin for 10 min at 800 × *g* at room temperature.
4. Discard the supernatant promptly and wash with 10 mL of ZPFM.
5. Spin for 5 min at 800 × *g* at room temperature.
6. Add 48 μL of the PCR product or 5 μg of linearized DNA into an electroporation cuvette.
7. Discard the supernatant promptly and resuspend cells at 1 × 10^8 cells/mL in ZPFM (e.g., 500 μL for one transfections).
8. Transfer 500 μL of cells to a 4 mm gap electroporation cuvette.
9. Electroporate cells twice with a Bio-Rad Gene Pulse Xcell electroporator (exponential decay wave pulse, voltage 1.7 kV, capacitance 25 μF, resistance ∞, cuvette 0.4 cm).
10. Transfer the contents of each cuvette to a 25 cm^2 flask containing 4.5 mL of prewarmed SDM79.
11. Leave the flasks overnight (typically 16 h) in a 28°C incubator. Then add only the selection drug for the transfected construct to the flask in the next morning. Incubate at 28°C. Passage the cells when the cell density gets high or the color of the culture gets yellow. Repeat until drug-resistant populations recover. Once the cells start to show a normal growth rate, add all the selection drugs. This "population" culture contains cells with one allele tagged with YFP.

3.4 Isolating Clones to Prepare Homozygously YFP-Tagged Strain

1. In the afternoon, 1 day before transfection, prepare a 10 mL of the population culture (per transfection) of 2 × 10^6 cells/mL with selection drugs. Prepare an extra 20 mL cell culture of 2 × 10^6 cells/mL without any drugs, which will be used to prepare conditioned media to isolate clones.
2. Transfect another YFP-tagging construct that has a different selection marker (see Sect. 3.3).
3. In the next day, isolate clones by limited dilution as follows to screen cells that have both alleles tagged with YFP.
4. Take the culture for conditioned media from the incubator. This culture should be overgrown. Spin for 5 min at 800 × *g* at room temperature. After centrifugation, take the supernatant and use as conditioned media. Conditioned media can be kept at −20°C. Use the parental strain of the transfection to make conditioned media.
5. Mix the following contents into 55 mL reagent reservoirs: 1 mL, 0.2 mL, or 40 μL of transfected culture (population), 15 mL of fresh SDM79, 5 mL of conditioned media, and 20 μL

of selection drugs (G418 and hygromycin) to make 20 mL of 1:20, 1:100, and 1:500 dilution cultures, respectively.

6. Plate 200 μL per well on 96-well flat-bottom plates. After plating, incubate the plates in a humidified incubator or in a sealed plastic container with wet tissue to avoid evaporation from the plate. Wait until drug-resistant cells grow. Typically, it takes 10 days to 2 weeks until clonal cells grow to a good density.
7. Once wells become yellow, select six to ten wells from the plate with the highest dilution factor. Transfer the 200 μL of cells to a 24-well plate and add 1 mL of fresh SDM79 with selection drugs. From this point, cells can be kept with all the drugs (puromycin, G418, and hygromycin). Incubate the plate in humid condition at 28°C for 2 days.
8. After 2 days, use 200 μL to check the genotype (see Sect. 3.5). If cells have the expected genotype, transfer them to a 25 cm^2 flask containing 5 mL of SDM79 and use for the deGradFP plasmid transfection.

3.5 Preparation of Cell Lysate for PCR

1. Take 200 μL from 0.5–2 × 10^7 cells/mL culture.
2. Spin 2 min at 800 × *g*.
3. Throw away the supernatant.
4. Add 120 μL of 50 mM NaOH 2 mM EDTA.
5. 95°C for 10 min.
6. Add 24 μL of 450 mM Tris-HCl pH 6.8.
7. Use 0.5 μL for 10 μL PCR reaction.

3.6 PCR-Based Genotyping and Validation of Homozygous Tagging (Fig. 2)

Check if primers are designed just outside of the homology arms (Fig. 2a, black arrows).

1. Mix 25 μL of 2× PCR master mix, 1.5 μL of 10 μM forward primer, 1.5 μL of 10 μM reverse primer, and 22 μL of water.
2. Aliquot 10 μL in a 0.2 mL PCR tube and add 0.5 μL of cell lysate in each tube.
3. Set thermal cycler: Step 1: 98°C for 30 s. Step 2: 98°C for 10 s. Step 3: 55°C for 15 s. Step 4: 72°C for 1 min. Step 5: 30 cycles of steps 2–4. Step 6: 72°C for 30 s. Step 7: keep at 16°C.
4. Mix the whole PCR product with loading buffer and run it on a 1.5% agarose gel. Use a parental strain as a control.

3.7 Agarose Gel Electrophoresis to Check PCR Products

1. Make 1% (w/v) agarose in 1× TAE buffer (40 mM Tris base, 20 mM acetic acid, and 1 mM EDTA) in a microwave-safe container.
2. Heat in a microwave until agarose is completely dissolved.
3. Cool until you can touch it (around 50°C).

4. Add DNA dye (such as ethidium bromide or SYBRSafe), pour into a tray, and insert a comb.
5. Leave at room temperature until it becomes a gel.
6. Put the gel into the tank and pour 1× TAE buffer until it covers the gel.
7. Load PCR samples.
8. Run at constant voltage (e.g. 160 V for 20 min) and visualize with a transilluminator (expected bands are shown in Fig. 2b).

3.8 Transfection of deGradFP Construct

Select pBA2675 or pBA2705 depending on the protein localization. Digest 5 μg of the plasmid with *Not*I, purify by ethanol precipitation, and then transfect into the cell (see Sect. 3.3). Use phleomycin for the selection. Keep the population.

3.9 Induce Expression of deGradFP Construct

Once the population of drug-resistant cells recovers, add all the selection drugs. Two days before the deGradFP experiment, dilute cells into fresh media without any drugs. Induce deGradFP expression with a final concentration of 1 μg/mL doxycycline. Use the uninduced culture as a control. Check the depletion efficiency using fluorescence microscopy and/or Western blotting. An example is shown in Fig. 3.

4 Notes

1. Both deGradFP and AID recruit an endogenous ubiquitin ligase to a target protein and deplete it through the ubiquitin-proteasome system. deGradFP requires expression of an F-box of the SCF ubiquitin ligase subunit fused with a GFP nanobody to induce protein depletion. In theory, this is slower than AID, which only requires the binding of a small molecule, auxin. The success rate is around 60% in the original deGradFP system in *Drosophila melanogaster* [13, 14]. In *T. brucei*, we got a similar success rate (Table 1). The success rate of AID was also not necessarily high. In our hands, deGradFP worked better for some proteins, while AID worked better for some other proteins. Some proteins were resistant to both approaches. Usually, we do not know why deGradFP, AID, or RNAi do not work for some proteins. It is therefore important to try multiple methods to determine the most effective method.
2. We regularly use EYFP, but GFP or Venus should also work because they are recognized by the GFP nanobody, VhhGFP4 [15, 16].

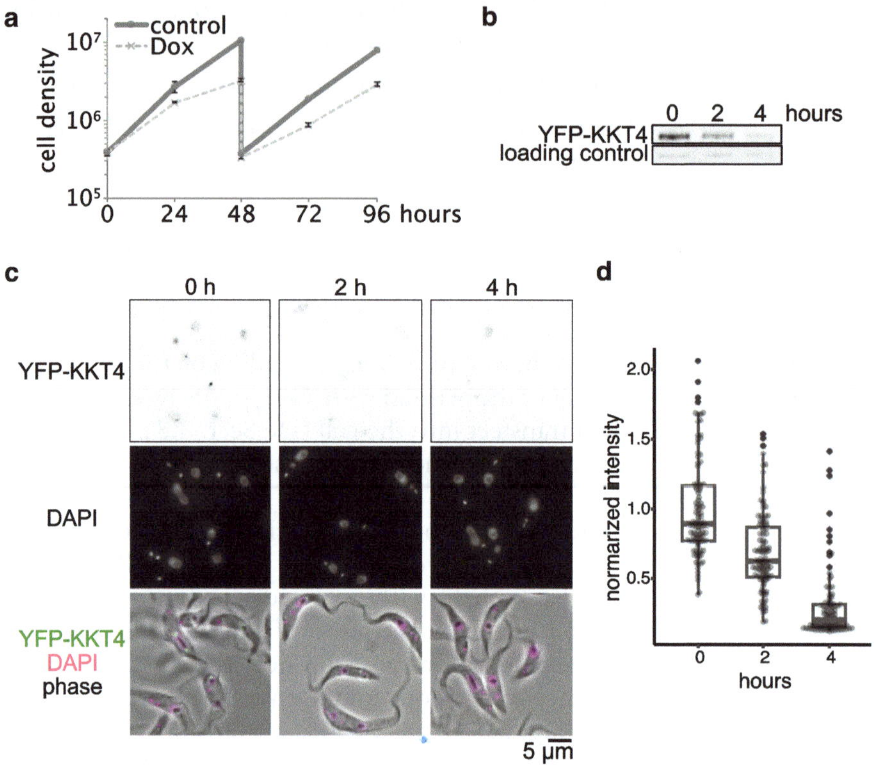

Fig. 3 YFP-KKT4 deGradFP. The homozygously YFP-tagged KKT4 (Tb927.8.3680) cell line was made by CRISPR techniques described in [6, 9]. Upstream forward primer (BA2510: GAATATTTTCTTTTTTTTTTGCTTA-CGCTTTCTATTTTAGTTACACAACAGCAAAAGGGGGCTAGATCAGAGGAATAAGGgtataatgcagacctgctgc), upstream reverse primer (BA2511: AGTGCTAAAATTCCCTGCATCTGAGTAATCAGCTGGGGATTAGATGCGAGTTGCTGTAGCA CAGACAGTGGATTGTCTCCCATactacccgatcctgatcc), and upstream sgRNA primer (BA4003: gaaattaatacgact-cactataggGGGGCTAGATCAGAGGAATAgttttagagctagaaatagc) were used to tag N-terminus of KKT4 in PCF1339. Forward primer (BA284: gatcgatc gcggccgcGATTCCAAAGGGAAGGAAGG) and reverse primer (BA283: gatcgatcgcggccgcCTAGACCCACCCCTCCAGGG) were used for genotyping to confirm homozygous tagging. The untagged KKT4 gene produces a 0.5 kb band; YFP-tagged genes produce around 2.6 kb for G418 marker and 2.8 kb for hygromycin marker. deGradFP was induced with 1 μg/mL doxycycline. Control was an uninduced culture. (**a**) Growth curve of YFP-KKT4 deGradFP. Error bars in the growth curve are standard deviation ($N = 3$). (**b**) Western blotting of samples collected every 2 h after induction of deGradFP. YFP-KKT4 was detected with primary antibody anti-GFP (TP401, OriGene) and secondary antibody IRDye anti-rabbit 800 (LI-COR). A loading control (PFR2) was detected with primary antibody L8C4 [17] and secondary antibody IRDye anti-mouse 680 (LI-COR). (**c**) Cells for microscopy imaging were fixed with formaldehyde and imaged as described in [6]. (**d**) Images were analyzed in ImageJ/Fiji [18]. YFP-KKT4 signal was measured inside the nucleus. DAPI channel was used to segment the nucleus by default threshold and then by Analyze Particle tool with size filter 2.5–20 μm^2. Nuclear regions were transferred by ROI Manager tool to YFP channel, and YFP signal was measured in each nuclear region. Integrated density was used to estimate the total protein signal. The mean value at 0-h time point was used to normalize the data. A total of 100 cells were analyzed at each time point

Table 1
List of successful and unsuccessful protein depletion by deGradFP in *T. brucei*

	Better than RNAi	Similar to RNAi	No clear growth defect (less effective than RNAi)
Nuclear protein	KKT3-YFP	YFP-KKT2	YFP-KKT16
	YFP-KIN-B	YFP-KKT17	KKT16-YFP
	YFP-KKT4	CPC1-YFP	YFP-KKT23
	YFP-AUK1		YFP-KKT18
			YFP-KKT14
			YFP-KKT15
			KIN-A-YFP
			YFP-KKT1
			YFP-KKT7
Cytoplasmic protein	SEC31-YFP		YFP-FLAM3

Nuclear proteins were transfected with pBA2675 (inducible expression of FBP75$^{1-200}$-NLS-VhhGFP4), while cytoplasmic proteins were transfected with pBA2705 (inducible expression of FBP75$^{1-200}$-VhhGFP4)

3. Use any cell lines, YFP-tagging systems, and an electroporation system available in your laboratory.
4. The restriction enzyme buffer for *Not*I usually contains enough salt to precipitate DNA. Add 1/10 volume of 5 M NaCl or 1/10 volume of 3 M sodium acetate pH 5.3 if your buffer does not contain enough salt to precipitate DNA.
5. Alternatively, two constructs with two different selectable marker genes can be transfected in one transfection step. The CRISPR-Cas9 system allows more efficient isolation of homozygously tagged clones in one step [6, 9].

Acknowledgments

This work was supported by a Wellcome Discovery Award (227243/Z/23/Z to B.A.).

References

1. Kim H-S, Li Z, Boothroyd C, Cross GAM (2013) Strategies to construct null and conditional null *Trypanosoma brucei* mutants using Cre-recombinase and loxP. Mol Biochem Parasitol 191:16–19. https://doi.org/10.1016/j.molbiopara.2013.08.001
2. Rico E, Jeacock L, Kovářová J, Horn D (2018) Inducible high-efficiency CRISPR-Cas9-targeted gene editing and precision base editing in African trypanosomes. Sci Rep 8:7960. https://doi.org/10.1038/s41598-018-26303-w

3. Merritt C, Stuart K (2013) Identification of essential and non-essential protein kinases by a fusion PCR method for efficient production of transgenic *Trypanosoma brucei*. Mol Biochem Parasitol 190:44–49. https://doi.org/10.1016/j.molbiopara.2013.05.002
4. Alsford S, Turner DJ, Obado SO et al (2011) High-throughput phenotyping using parallel sequencing of RNA interference targets in the African trypanosome. Genome Res 21:915–924. https://doi.org/10.1101/gr.115089.110
5. Ngô H, Tschudi C, Gull K, Ullu E (1998) Double-stranded RNA induces mRNA degradation in *Trypanosoma brucei*. Proc Natl Acad Sci U S A 95:14687–14692. https://doi.org/10.1073/pnas.95.25.14687
6. Ishii M, Akiyoshi B (2022) Targeted protein degradation using deGradFP in *Trypanosoma brucei*. Wellcome Open Res 7:175. https://doi.org/10.12688/wellcomeopenres.17964.2
7. Klein A, McDowell A, Reis CRS et al (2025) An auxin-inducible degron system for trypanosomes. bioRxiv. https://doi.org/10.1101/2025.06.17.660065
8. Poon SK, Peacock L, Gibson W et al (2012) A modular and optimized single marker system for generating *Trypanosoma brucei* cell lines expressing T7 RNA polymerase and the tetracycline repressor. Open Biol 2:110037. https://doi.org/10.1098/rsob.110037
9. Beneke T, Madden R, Makin L et al (2017) A CRISPR Cas9 high-throughput genome editing toolkit for kinetoplastids. R Soc Open Sci 4:170095. https://doi.org/10.1098/rsos.170095
10. Paterou A, Sáez Conde J, Týč J et al (2025) A comprehensive toolkit for protein localization and functional analysis in trypanosomatids. Open Biol 15:240361. https://doi.org/10.1098/rsob.240361
11. Kelly S, Reed J, Kramer S et al (2007) Functional genomics in *Trypanosoma brucei*: a collection of vectors for the expression of tagged proteins from endogenous and ectopic gene loci. Mol Biochem Parasitol 154:103–109. https://doi.org/10.1016/j.molbiopara.2007.03.012
12. Wirtz E, Leal S, Ochatt C, Cross GAM (1999) A tightly regulated inducible expression system for conditional gene knockouts and dominant-negative genetics in *Trypanosoma brucei*. Mol Biochem Parasitol 99:89–101. https://doi.org/10.1016/S0166-6851(99)00002-X
13. Caussinus E, Kanca O, Affolter M (2012) Fluorescent fusion protein knockout mediated by anti-GFP nanobody. Nat Struct Mol Biol 19:117–121. https://doi.org/10.1038/nsmb.2180
14. Caussinus E, Affolter M (2016) deGradFP: a system to knockdown GFP-tagged proteins. In: Dahmann C (ed) Drosophila: methods and protocols. Springer, New York, NY, pp 177–187
15. Saerens D, Pellis M, Loris R et al (2005) Identification of a universal VHH framework to graft non-canonical antigen-binding loops of camel single-domain antibodies. J Mol Biol 352:597–607. https://doi.org/10.1016/j.jmb.2005.07.038
16. Rothbauer U, Zolghadr K, Muyldermans S et al (2008) A versatile nanotrap for biochemical and functional studies with fluorescent fusion proteins. Mol Cell Proteomics 7:282–289. https://doi.org/10.1074/mcp.M700342-MCP200
17. Kohl L, Sherwin T, Gull K (1999) Assembly of the paraflagellar rod and the flagellum attachment zone complex during the *Trypanosoma brucei* cell cycle. J Eukaryot Microbiol 46:105–109. https://doi.org/10.1111/j.1550-7408.1999.tb04592.x
18. Schneider CA, Rasband WS, Eliceiri KW (2012) NIH image to ImageJ: 25 years of image analysis. Nat Methods 9:671–675. https://doi.org/10.1038/nmeth.2089

Part V

Methods for Studying Bacterial Degrons

Chapter 15

Production of Purified Proteins Tagged with ssrA-Derived Degrons and Their Degradation by the *Escherichia coli* ClpXP System

Maria M. Klimecka, Matylda A. Izert-Nowakowska and Maria W. Górna

Abstract

The ssrA degron is widely employed in fusion proteins to regulate protein stability in bacteria or to serve as an interaction module. These uses take advantage of the ssrA tag's dual functionality—its ability to bind the SspB adaptor and to recruit the ClpXP protease. A comprehensive comparison of these functions across a standardized set of degron variants can be helpful in the development of variants specifically optimized for particular applications and their use in targeted protein degradation systems. Here, we describe a simple and efficient purification protocol for a set of ssrA-based degrons and a degradation assay that can be used as the primary assay to compare the effectiveness of degradation of degron-tagged proteins.

Keywords Protease, Degradation, Degron, ssrA, ClpXP, SspB

1 Introduction

Degrons are often used in various directed protein degradation methods as a part of PROteolysis TArgeting Chimeras (PROTACs) responsible for binding them to degradation system components [1, 2]. In bacteria, one of the best described degrons and the most frequently used in biotechnology is the ssrA tag sequence and its variants [3–5]. In the bacterial quality control pathway, ssrA tag is added in the trans-translation process during which the tmRNA (transfer-messenger RNA) enters stalled ribosomes, replaces the mRNA, and promotes the translation of a peptide tag (ssrA) encoded within tmRNA onto the C terminus of the nascent polypeptide [6].

The ssrA tag is highly conserved among bacterial species and, in *Escherichia coli*, it is primarily recognized by the ClpXP protease, a member of the AAA+ (ATPases Associated with diverse cellular Activities) family. The ClpX component functions as an unfoldase and forms a hexameric ring that features dimers of the N-terminal zinc-binding domain (ZBD). The associated peptidase, ClpP,

Maria W. Górna et al. (eds.), *Degrons*, Methods in Molecular Biology, vol. 3069,
https://doi.org/10.1007/978-1-0716-5508-5_15,

assembles into a double heptameric ring capable of interacting with ClpX hexamers on one or both faces of the ClpP barrel. The ssrA tag itself is an 11-amino-acid sequence (AANDENYALAA), where the AANDENY segment mediates binding to the ClpX adaptor protein SspB, and the ALAA sequence interacts directly with the central pore of the ClpX unfoldase [7]. Due to this modular structure, the ssrA tag is often utilized in various biotechnological applications.

When designing bacterial PROTACs (BacPROTACs), the ability to simultaneously compare the different properties of the ssrA tag and its versions in one experimental setup is very helpful, especially when it comes to comparing degradation efficiency. In the previously published work [8, 9], we included both in vitro and in vivo methods for studying binding and degradation of eGFP-degrons, and the protocols outlined here will enable to obtain eGFP proteins with ssrA tag-based degrons in vitro and compare their degradation efficiency.

2 Materials

2.1 eGFP-Degrons Expression and Purification

1. Media: LB Lennox agar, LB Lennox.
2. Antibiotics: carbenicillin.
3. Strains: *ΔclpP* Keio collection strain (JW0427).
4. Expression inductor: arabinose, powder.
5. Incubator with shaking.
6. Centrifuges: for big volumes (cultures, ~1 l) for centrifugation at 5180 × *g*; for small volumes (lysates, ~50 ml) for centrifugation at 48,380 × *g* and 5000 × *g*, all at 4°C.
7. Buffers: Lysis/loading buffer: 50 mM Tris pH 8.0, 20 mM imidazole, 500 mM NaCl.
   - Elution buffer: loading buffer with 1 M imidazole.
   - SEC buffer: 25 mM HEPES pH 7.6, 200 mM KCl, 5 mM $MgCl_2$, 10% glycerol.
8. Lysozyme, powder.
9. DNase (20 U).
10. FPLC columns: Ni^{2+}-affinity column 5 ml, preparative (for tandem configuration) or analytical (for separate purification steps) size-exclusion chromatography column (SEC) (with a broad fractionation range from MW 10 to ~600 kDa).
11. FPLC system.
12. Rotator.
13. Sonicator, for example, Vibra-cell VCX130 (Sonics & Materials Inc.).

14. Sample filtration: 20 ml syringe, syringe filters 0.8 μm.
15. 10 kDa MWCO concentrators.
16. Automatic pipettes: 2–10 μl, 20–200 μl, 100–1000 μl, and proper tips.
17. 50 ml, 15 ml, and 1.5 ml tubes.
18. Thermomixer.
19. Tris–glycine SDS-PAGE gels (12%).
20. Tris–glycine SDS-PAGE loading buffer.
21. Tris–glycine SDS-PAGE running buffer.
22. Coomassie staining reagent.
23. SDS-PAGE apparatus.

2.2 In Vitro Degradation

1. Reaction buffer: 50 mM Tris–HCl pH 7.5, 200 mM NaCl, 5 mM $MgCl_2$.
2. *E. coli* degradation system proteins: ClpX, ClpP, sspB.
3. Various eGFP-degrons including eGFP-ssrA as a positive control.
4. eGFP as a negative control.
5. ATP solution.
6. Creatine phosphate.
7. Creatine kinase.
8. Automatic pipettes: 0.5–2 μl, 2–10 μl, 20–200 μl, and proper tips.
9. 1.5 ml tubes.
10. Thermomixer.
11. Tris–glycine SDS-PAGE gels (12%).
12. Tris–glycine SDS-PAGE loading buffer.
13. Tris–glycine SDS-PAGE running buffer.
14. Coomassie staining reagent.
15. SDS-PAGE apparatus.

3 Methods

3.1 eGFP-Degrons Expression and Purification

For in vitro studies of eGFP-degrons, expression in bacterial system and purification of recombinant eGFP-degrons is necessary. Constructs of all studied eGFP-degrons can be obtained as previously described in Klimecka et al. (2021) [9], as N-His-TEV-eGFP-degrons (i.e., fusion with an N-terminal purification tag and a C-terminal degron). For efficient expression without visible protein degradation, the use of pBAD vector as the backbone vector and the *ΔclpP* Keio collection strain (JW0427) as the expression strain is recommended (*see* **Note 1**)(Fig. 1A, B). All steps of purification should be carried out at 4°C (on ice) to limit spontaneous protein degradation. The eGFP-degrons can be purified on Ni^{2+}-affinity and SEC columns in a tandem configuration without any purification tag (His-TEV) removal (*see* **Note 2**). *E. coli* degradation system proteins (ClpX, ClpP, and SspB) were obtained as in [9].

3.1.1 eGFP-Degrons Expression

1. Transfer one *E. coli* colony from a carbenicillin LB agar plate to 20 ml of liquid LB media containing 100 mg/ml carbenicillin and culture overnight at 37°C with shaking (150–200 rpm).
2. The next day, transfer the whole 20 ml culture (dilution 1:50) to 1 l of fresh LB media with carbenicillin (100 mg/ml). For induction, add arabinose to 0.02% concentration to the culture at OD ~0.5. After overexpression at 30°C for 3 h, 140 rpm, collect bacterial pellet by spinning cultures at 5000 rpm (5180 × *g*) for 30 min at 4°C. Remove media, transfer pellet to 50 ml falcon tube, and freeze at −20°C.

3.1.2 eGFP-Degrons Purification

1. Resuspend frozen bacterial pellet in ~30 ml of Lysis buffer (*see* **Note 3**). To start lysis, add lysozyme (100 mg) and DNase (~20 U) and incubate the lysate at 4°C with gentle mixing on a rotator for 30 min.
2. For thorough lysis, use sonication (15 min, 45 s on/15 s off, 40% amplitude, on ice) (*see* **Note 4**).
3. Next, clear the lysate by centrifugation: 20,000 rpm (48,380 × *g*), 30 min, 4°C.
4. To avoid aggregates, filter the cleared lysate additionally using syringe filters (0.8 μm).

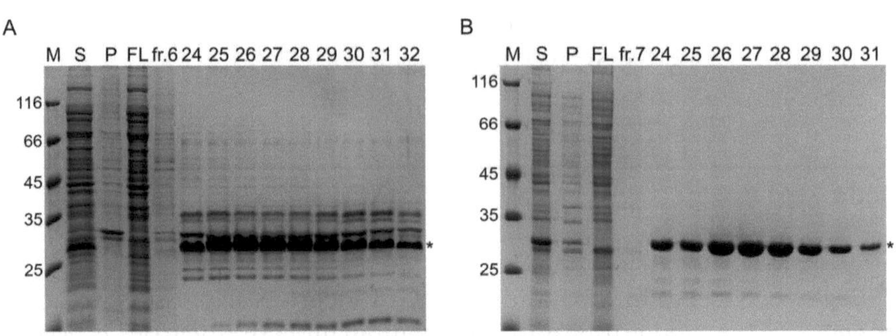

Fig. 1 Comparison of purification of eGFP-SENY-ADAS between (**A**) expression in BL21 DE3 and (**B**) expression in *ΔclpP* strain using tandem configuration of Ni^{2+}-affinity and SEC column; M—protein mass marker; S—cleared bacterial lysate loaded on affinity column; P—bacterial pellet; FL—flow-through from affinity column; 6(7); 24-31(32)—individual fractions eluted from the SEC column; * —intact, full-length eGFP-SENY-ADAS

5. Apply the lysate on an FPLC system equilibrated with the lysis and elution buffer. As a default, a sequential use of a Ni^{2+}-affinity column and an SEC column is described. For other options, *see* **Note 5**.
6. The standard conditions, for example, flow rate specified by the manufacturer of the column, can be used for affinity chromatography. For further tips, *see* **Note 6**. Of all the protein fractions collected during elution, those visible on the chromatogram with increased absorbance signal at 280 nm (*see* **Note 7**) should be analyzed by SDS-PAGE. In addition, during the initial optimization of the purification, SDS-PAGE analysis of the different steps of preparing the lysate before applying it to the column will be beneficial (*see* **Note 8**).
7. Run SDS-PAGE (*see* **Note 9**).
8. Fractions identified on the SDS-PAGE gel as containing mainly the eGFP-degron should be combined and then concentrated at 4°C using 10 kDa MWCO concentrators according to their manufacturer's instructions (*see* **Note 10**) to the volume recommended by the SEC column manufacturer depending on what type of the column is used.
9. In order to remove potential aggregates, the sample should be additionally centrifuged (~21,000 × *g*, ~10 min, 4°C).
10. Using 1 ml syringe, apply protein sample on 1 ml loop connected to FPLC system with the SEC column.
11. Fractions with an increased absorbance signal at 280 nm should be analyzed by SDS-PAGE.
12. Run SDS-PAGE.
13. Fractions selected on the basis of the FPLC chromatogram and SDS-PAGE analysis should be combined, and then the protein concentration should be measured.
14. Freeze the obtained protein in liquid nitrogen and store at −80°C (Fig. 2A, B).

3.2 SDS-PAGE Analysis of In Vitro Degradation of eGFP-Degrons

For degradation experiments, the full ClpX/ClpP/SspB degradation system can be used as described previously in Klimecka et al. [9].

Addition of creatine phosphate and creatine kinase to the reaction mixture is required for efficient ATP recovery throughout the degradation reaction (*see* **Note 11**).

3.2.1 In Vitro Degradation Sample Preparation and Analysis

1. For convenience, perform in vitro degradation of eGFP-degrons by ClpXP in the 100 μl final volume in 1.5 ml tubes. To prepare reaction mixtures, pipette proper amount of the reaction buffer (50 mM Tris-HCl pH 7.5, 200 mM NaCl, 5 mM $MgCl_2$), 0.08 μM $ClpX_6$, eGFP-degron (1 μM), 2.5 μM SspB (if added), 4 mM ATP, 2.5 mM creatine phosphate, and 50 μg/ml

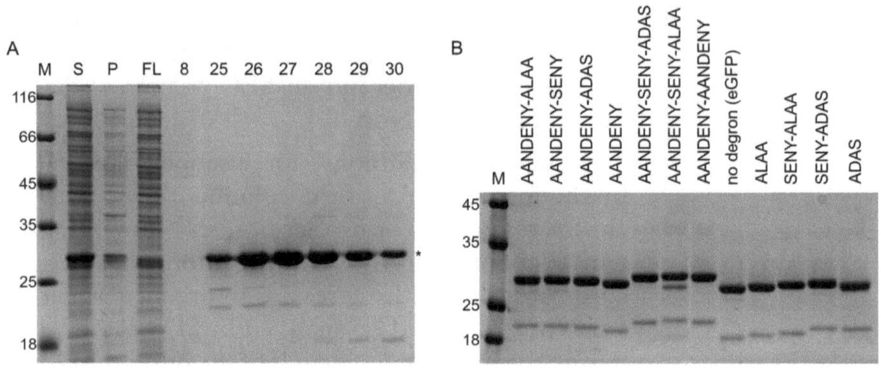

Fig. 2 Purification of eGFP-degrons. (**A**) SDS-PAGE gel with fractions collected after tandem purification of HisTEV-eGFP expressed in *E. coli ΔclpP* strain (JW0427). M—protein mass marker; S—cleared bacterial lysate before loading on the Ni^{2+}-affinity column; P—bacterial pellet; FL—flow-through from the Ni^{2+}-affinity column; 8-30—Individual fractions eluted from the SEC column. *—Intact, full-length eGFP. (**B**) SDS-PAGE gel with all obtained eGFP-degrons

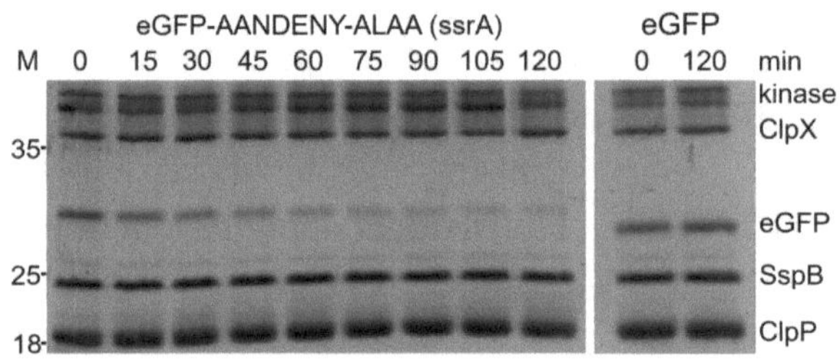

Fig. 3 SDS-PAGE analysis of the in vitro eGFP-ssrA degradation by ClpXP. The progress of eGFP-ssrA degradation in the presence of 2.5 μM $SspB_2$ protein was sampled over a time course of 2 h (left). Degradation of eGFP without a degron tag served as a control (right). M—protein mass marker

creatine kinase and incubate at 30°C for 30 min. For use of proper controls, *see* **Note 12**.

2. Start the degradation reaction by the addition of 0.21 μM $ClpP_{14}$ for the final protease machinery assembly in the $ClpX_6$:$ClpP_{14}$ = 3:8 ratio and continue incubation at 30°C.
3. Collect 10 μl samples in convenient time intervals up to 120 min. Each time stop degradation in the sample by adding the SDS-PAGE loading buffer and incubate for 5 min at 95°C.
4. Apply the prepared samples (10 μl) on 12% SDS-PAGE gels and develop at 200 V.
5. Visualize resolved proteins using Coomasie staining reagent (Fig. 3).

4 Notes

1. The vector used for eGFP-degron cloning should provide efficient but precisely regulated expression. Such a vector could be, for example, pBAD, which has a highly inducible araBAD operon system responsible for very tightly controlled expression. Additionally, the use of *ΔclpP* strain from Keio collection (JW0427) is highly recommended because spontaneous degradation by the endogenous ClpXP complex can be observed when expressed in another strain lacking genes for L-arabinose catabolism. This was the case for the expression of eGFP-degrons in the *E. coli* BL21 DE3 strain, where, regardless of the conditions used for expression and purification, significant degradation of the eGFP-degrons was observed already at the beginning of purification (Fig. 1).
2. Removal of His tag is not necessary due to the fact that it is located at the N-terminus, far from the degrons themselves, and does not affect degradation.
3. It is good practice to match the amount of buffer proportional to the mass of the bacterial pellet obtained during the expression. To make the subsequent lysis efficient, usually no more than 30–40 ml buffer should be used to suspend the pellet obtained from 1 l of culture.
4. During sonication, it is advisable to use pause to avoid overheating the sample. This can be an interval of 30 s on and 30 s off but also 45 s on and 15 s off, both were sufficient to keep the lysate cool.
5. An FPLC system having at least two column valves allows eGFP-degron purification in tandem mode. This relies on the fact that the proteins eluted from the Ni^{2+}-affinity column can be immediately applied to the SEC column without any additional concentration, and the presence of a fluorescent eGFP makes it easy to track the sample throughout purification. After pre-optimization, this method saves time especially when purifying multiple eGFP-degrons.
6. Typically, the binding of proteins with a His tag to the Ni^{2+}-affinity column is quite fast and allows the use of flow rates of, for example, 2 ml/min (the recommended flow rate should be stated in the manufacturer's manual) throughout the FPLC run (at 4–10°C); however, the application of the sample to the column should be somewhat slower to maximize the binding efficiency (e.g., 1 ml/min).
7. The use of additional wavelengths can be useful. At 495 nm, for example, an increase in UV absorbance signal for eGFP can be observed.

8. By analyzing additional samples on SDS-PAGE, it can be determined whether the proteins are soluble (comparing the cleared, filtered lysate with the insoluble pellet after lysis and centrifugation) and whether all the protein has bound to the column (protein should be then absent in the flow-through sample).
9. 12% Tris–glycine gels are a good choice but 10% or 15% will also be suitable, either precast or self-made. Voltage during the run can be up to 240 V for 10% and 12% gels and 210–220 V for 15% gels.
10. During concentration between successive centrifugations, it is advisable to mix the sample by pipetting to avoid a local increase in protein concentration and its potential precipitation.
11. Phosphoenolpyruvate (PEP) together with pyruvate kinase (PK) can also be used as an alternative ATP regeneration system.
12. Degron-less eGFP should be used as a negative control. Additionally, the reaction mixture without ClpP should be used as an additional control to check whether eGFP-degrons undergo spontaneous degradation during incubation.

References

1. Verma R, Mohl D, Deshaies RJ (2020) Harnessing the power of proteolysis for targeted protein inactivation. Mol Cell 77:446–460. https://doi.org/10.1016/j.molcel.2020.01.010
2. Harris TJ, Trader DJ (2025) Exploration of degrons and their ability to mediate targeted protein degradation. RSC Med Chem 16:1067–1082. https://doi.org/10.1039/d4md00787e
3. Saunders RA, Stinson BM, Baker TA, Sauer RT (2020) Multistep substrate binding and engagement by the AAA+ ClpXP protease. Proc Natl Acad Sci U S A 117:28005–28013. https://doi.org/10.1073/pnas.2010804117
4. Fei X, Bell TA, Barkow SR, Baker TA, Sauer RT (2020) Structural basis of ClpXP recognition and unfolding of ssra-tagged substrates. eLife 9:1–39. https://doi.org/10.7554/eLife.61496
5. Izert MA, Klimecka MM, Górna MW (2021) Applications of bacterial degrons and degraders — toward targeted protein degradation in bacteria. Front Mol Biosci 8:669762. https://doi.org/10.3389/fmolb.2021.669762
6. Keiler KC (2008) Biology of trans-translation. Annu Rev Microbiol 62:133–151. https://doi.org/10.1146/ANNUREV.MICRO.62.081307.162948
7. Baker TA, Sauer RT (2012) ClpXP, an ATP-powered unfolding and protein-degradation machine. Biochim Biophys Acta 1823:15–28. https://doi.org/10.1016/J.BBAMCR.2011.06.007
8. Izert-Nowakowska MA, Klimecka MM, Antosiewicz A, Wróblewski K, Kowalski JJ, Bandyra KJ et al (2025) Targeted protein degradation in Escherichia coli using CLIPPERs. EMBO Rep 26:3994. https://doi.org/10.1038/S44319-025-00510-9
9. Klimecka MM, Antosiewicz A, Izert MA, Szybowska PE, Twardowski PK, Delaunay C et al (2021) A uniform benchmark for testing Ssra-derived degrons in the *Escherichia coli* ClpXP degradation pathway. Molecules 26:5936. https://doi.org/10.3390/molecules26195936

Chapter 16

Evaluation of Degron Motifs in *Escherichia coli* Using a Fluorescent Reporter

Matylda A. Izert-Nowakowska, Patrycja E. Szybowska, Maria M. Klimecka and Maria W. Górna

Abstract

Fluorescent reporters provide a useful tool for studying degron motifs. Fusing a degron of interest to a fluorescent protein allows to accurately track protein levels overtime to characterize the degradation kinetics of studied degrons. Here, we describe a rapid and simple method to study degron peptides in *Escherichia coli* using plasmid-encoded eGFP-degron fusion constructs. The described methods provide an accessible workflow to evaluate degrons. We provide protocols for generation of pBAD plasmids encoding the studied constructs and two different methods for evaluating degrons: an end-point fluorescence measurement on agar plates and a kinetic measurement in liquid cultures in a 96-well format for high-throughput degron studies.

Keywords ssrA degron, GFP, Fluorescent reporter, Plate test

1 Introduction

Degron motifs in bacteria determine protein turnover by regulating protein stability [1]. They drive protein degradation, allowing bacteria to maintain the correct expression levels of essential proteins in response to external stimuli. This allows the bacteria to quickly respond to environmental stress and adapt to the changing conditions. Fluorescent reporter assays are powerful tools to study peptide degrons and their ability to drive protein degradation [2]. Isolating a degron motif and fusing it to a non-native fluorescent protein allows for studying specific degrons in their cellular context. This can be beneficial for identification of new degradation motifs or evaluating engineered degrons for their use in synthetic applications [3, 4]. Here, we present a rapid and simple screening method using a plasmid-encoded fluorescent reporter for studying peptide degrons in *Escherichia coli*. We provide protocols for two different degradation assays. The plate spot assay allows for rapid screening of

Maria W. Górna et al. (eds.), *Degrons*, Methods in Molecular Biology, vol. 3069,
https://doi.org/10.1007/978-1-0716-5508-5_16,

degron motifs efficiency. It can be used as initial screening method before more thorough evaluation by time-dependent degradation assay. The assays can help to validate the degradation pathway by using single deletion mutant strains, which are easily accessible from Keio collection [5]. Alternatively, when protease driving the degradation is not known, the assay can be performed in presence of a broad-spectrum protease inhibitor bortezomib. The presented protocols do not require special setup or equipment and can be performed with just fluorescent gel scanner and plate reader, which are available in most laboratories, making them a powerful screening strategy before committing to more in-depth degron investigation in vitro [6], described separately in a dedicated chapter.

2 Materials

2.1 Cloning

1. Inducible plasmid encoding eGFP under arabinose promoter (Addgene plasmid #54762 EGFP-pBAD) (*see* **Note 1**).
2. Template-specific primers with overhangs encoding a degron sequence of interest.
3. High-fidelity DNA polymerase (e.g., Phusion DNA polymerase).
4. 10 mM dNTPs mix.
5. Nuclease-free water.
6. Thermal cycler.
7. DpnI enzyme.
8. Digestion buffer.
9. DNA loading dye.
10. DNA ladder.
11. Agarose (molecular biology grade).
12. TAE buffer (Tris base 40 mM, acetic acid 20 mM, EDTA 1 mM).
13. DNA dye (e.g., SYBR™ Safe).
14. Horizontal electrophoresis apparatus and power pack.
15. Clean scalpel or gel-cutting device.
16. Gel-out DNA purification kit or PCR DNA purification kit.
17. Nano-spectrophotometer (e.g., Nanodrop).
18. Polynucleotide kinase (PNK).
19. T4 DNA ligase.
20. DNA ligase buffer.
21. 10 mM ATP.
22. Electrocompetent *E. coli* strain compatible with cloning (DH5α or Top10).

23. Thermoblock.
24. Luria–Bertani (LB) agar plates with ampicillin (100 μg/mL).
25. Miniprep DNA isolation kit.

2.2 Bacterial Cultures for Degron Expression

1. Chemically competent *E. coli* strain BW25113 and protease mutant strains available from Keio strain collection [5].
2. Luria–Bertani (LB) medium.
3. 20% sterile L(+)-arabinose solution.
4. Ampicillin 100 mg/mL.
5. Kanamycin 15 mg/mL.
6. LB agar plates with 100 μg/mL ampicillin and 0.001% arabinose or with 100 μg/mL ampicillin, 15 μg/mL kanamycin, and 0.001% L(+)-arabinose.
7. Bortezomib stock solutions in DMSO at 50, 25, 10, 5, and 1 mM (optional).
8. Shaking incubator.
9. Spectrophotometer with compatible cuvettes.

2.3 Degron Evaluation

1. Ampicillin 100 mg/mL.
2. Kanamycin 15 mg/mL.
3. Spectinomycin 100 mg/mL.
4. M9 medium: Na_2HPO_4 6 g/L, KH_2PO_4 3 g/L, NaCl 0.5 g/L, NH_4Cl 1 g/L, $MgSO_4$ 120 mg/L, $CaCl_2$ 11 mg/L, $CaCl_2 \cdot 2\ H_2O$ 15 mg/L, $Na_2EDTA \cdot 2\ H_2O$ 30 mg/L, $FeCl_3 \cdot 6\ H_2O$ 25 mg/L, $CuSO_4 \cdot 5\ H_2O$ 2.4 mg/L, $MnSO_4 \cdot H_2O$ 1.8 mg/L, $ZnSO_4 \cdot 7\ H_2O$ 0.27 mg/L, $CoCl_2$ 0.27 mg/L, thiamine 10 mg/L, biotin 10 mg/L.
5. 94-well black plate with optic bottom.
6. Gel imager with optic filters for GFP (excitation at $\lambda = 488$ nm and emission at $\lambda = 520$ nm).
7. Microplate reader with shaking and temperature control with optical filters for GFP (excitation at $\lambda = 488$ nm and emission at $\lambda = 520$ nm).

3 Methods

3.1 Preparation of Plasmids the GFP-Degron Fusion Construct by Site-Directed Mutagenesis

1. Insertion of degron sequences up to 15–20 amino acids can be easily achieved by site-directed mutagenesis. Design the primers according to the orientation of your investigated degrons. If you do not know their orientation, create constructs with both N-terminal and C-terminal fusion to eGFP. When designing

your primers, aim for a similar hybridization temperature (ideally between 60°C and 65°C) for the parts which are complementary to the plasmid template. The plasmid overhangs should encode your degron and should face away from each other to amplify the entire plasmid while adding the degron in fusion with eGFP (Fig. 1). If you wish to investigate longer degron sequences, you might have to consider other cloning techniques (e.g., Gibson assembly).

2. Prepare a 50 μL PCR reaction containing:
   10 ng eGFP-pBAD plasmid template.
   200 μM dNTPs mix.
   0.5 μM forward primer.
   0.5 μM reverse primer.
   1× polymerase-specific reaction buffer.
   2 U high-fidelity DNA polymerase.
   Run the reaction in the thermal cycler (*see* **Note 2**).
   - Initial denaturation: 98°C, 30 s.
   – Repeat the following steps 30 times:
   - Denaturation: 98°C, 10 s.
   - Primer annealing: 65°C, 15 s.
   - Elongation: 72°C, 2 min.

   - Final elongation: 72°C, 5 min.

   Run a 5 μL aliquot of the PCR product on 0.8% agarose gel with DNA dye in 1× TAE buffer to evaluate the presence and purity of your PCR product. Visualize with a gel imager or transilluminator. The template plasmid has a size of 5429 bp and, for short degron peptide motifs, the size difference be-

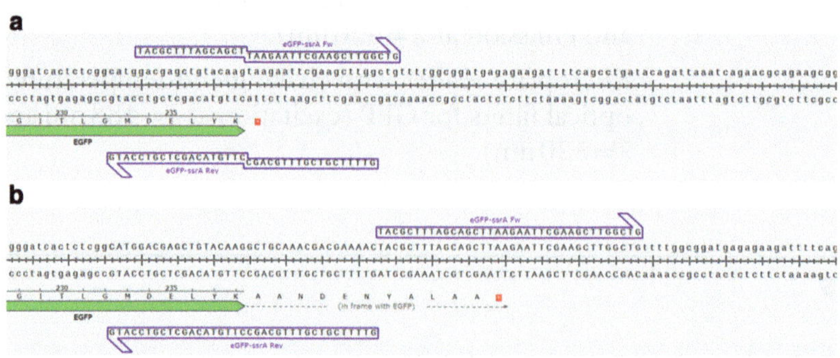

Fig. 1 An example of primer design for introducing ssrA degron at the C-terminus of eGFP in EGFP-pBAD plasmid. Primers are outlined with purple arrows. (**a**) A representation of the primer hybridization sites in the parental plasmid. Parts of forward and reverse primers hybridize to the vector and the overhangs encode ssrA degron. (**b**) A representation of the same plasmid region after introducing the ssrA degron by site-directed mutagenesis

tween modified and parental plasmid might not be distinguishable on the gel.

3. To remove the template DNA from the reaction mix, dilute the PCR product to 89 μL (so that the final reaction volume would be 100 μL) with nuclease-free water, add the enzyme-specific buffer to the final concentration of 1×, and add DpnI enzyme. Incubate the reaction at 37°C for 15–30 min, according to the manufacturer's instructions. Purify the DNA (*see* **Note 3**). Quantify the DNA concentration with a nano-spectrophotometer.
4. Assemble the circularization reaction with PNK to phosphorylate 5′ DNA ends and T4 ligase in a total volume of 20 μL:
   100 ng purified DNA.
   1× DNA ligase buffer.
   250 μM ATP (optional).
   5 U PNK.
   2.5 U T4 DNA ligase.
   Incubate the reaction for 1 h at room temperature. Alternatively, the reaction might be left overnight at 4°C.
5. Transform 5 μL of the ligation mixture to a 50 μL aliquot of chemically competent cloning-compatible *E. coli* strain. Grow bacteria on selection LB agar plates with ampicillin at 37°C overnight. The following day, inoculate a couple of colonies in 5 mL of LB with 100 μg/mL ampicillin and grow for 16–20 h at 37°C in a shaking incubator. Isolate plasmid DNA and confirm the insertion of the degron sequence by DNA sequencing.

3.2 Preparation of Bacterial Cultures for Degron Evaluation

Transform plasmids encoding eGFP-degron fusions in *E. coli* strain BW25113. Include control with unmodified degron-less eGFP construct and bacteria transformed with an empty vector (such as pBAD-6xHis-TEV, *see* **Note 4**). If you know the protease recognizing the investigated degron motifs, you can transform the plasmids in parallel to a single gene mutant strain (e.g., from Keio collection). Grow the bacteria overnight at 37°C on plates with 100 μg/mL ampicillin (wild-type BW25113) or 100 μg/mL ampicillin and 15 μg/mL kanamycin (mutant strains). Inoculate single colonies of transformed bacteria in 5 mL LB with appropriate selection antibiotics and grow overnight at 37°C in a shaker incubator. At this point, you can use aliquots of the cultures for a spot assay on agar plates (see below). For the dynamic degradation evaluation in liquid cultures, dilute the cultures 1:100 in 5 mL of LB with selection antibiotics and grow to mid-exponential phase (OD_{600} = 0.5–0.6). Induce the cultures with 0.005% L-arabinose and incubate overnight at 18°C with shaking.

3.3 Plate Spot Assay

Dilute the overnight cultures 1:100 in sterile LB. Place 3 μL drops on LB agar plates with selection antibiotics (100 μg/mL ampicillin

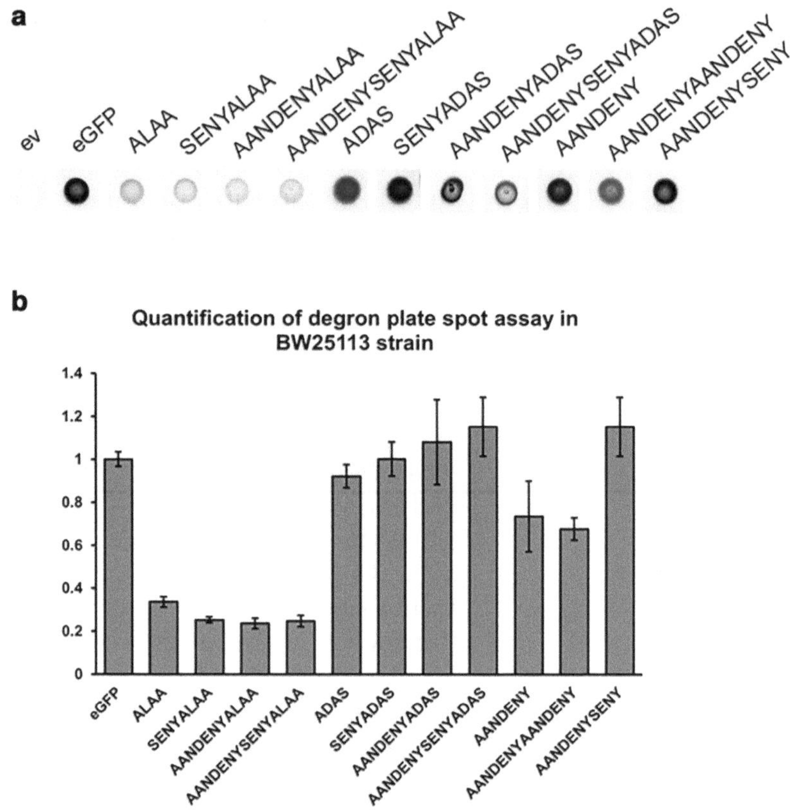

Fig. 2 An example of plate spot assay results in BW25113 strain [6]. (**a**) Pictures of bacterial spot colonies grown on LB agar in presence of 0.005% L(+)-arabinose expressing eGFP in fusion with peptide degrons. (**b**) Quantification of colony fluorescence. The bars represent mean values from one biological repeat performed in triplicate. The error bars represent SEM

for BW25113 strain and 100 μg/mL ampicillin with 15 μg/mL kanamycin for Keio mutant strains) supplemented with 0.001% L-arabinose. Incubate the plates overnight at 37°C. Image plates on a gel imager with optic filters for GFP or similar (excitation at $\lambda = 488$ nm and emission at $\lambda = 520$ nm, such as Alexa Fluor 488 or fluorescein) (Fig. 2).

3.4 Time-Dependent Microplate Degradation Assay

Prepare a 96-well black plate with an optical bottom. Assemble the plate on ice to reduce the degradation of eGFP-degron constructs during preparation of the plate. Pipette 190 μL of M9 medium with 100 μg/mL ampicillin (or 100 μg/mL ampicillin and 15 μg/mL kanamycin for measurements in mutant strains), 100 μg/mL spectinomycin, and 0.2% glucose to ensure the expression arrest of the constructs. Then add 10 μL of the induced cultures to the wells (to achieve 1:20 dilution) in triplicates. Place the plate in a microplate reader pre-heated to 30°C and measure the fluorescence and OD_{600}

in the wells at 15 min interval for 6 h with shaking in between the measurements (*see* **Note 5;** Fig. 3a–c).

3.5 Bortezomib Inhibition Assay

For bortezomib inhibition assay prepare the bacterial cultures as described above. When preparing the 96-well plate with M9 medium, add bortezomib in DMSO at concentrations 500, 250, 100, 50, 10 μ M, or 1% DMSO. Then, dilute the cultures 1:20 in the prepared microplate and measure in a microplate reader as described above (Fig. 3d).

3.6 Data Evaluation

Plate spot test can be quantified in image quantification software such as Fiji [7]. Select the region of interest (ROI) by drawing circles around the bacterial spots. Quantify the signal intensity by measuring brightness of the spots. To account for the background, subtract the signal of bacteria expressing the empty vector.

To evaluate the results from plate reader measurements, normalise the fluorescence value of each datapoint by dividing the fluorescence value by OD_{600} value. Subtract the average value obtained for bacteria expressing empty vector from data points at each time

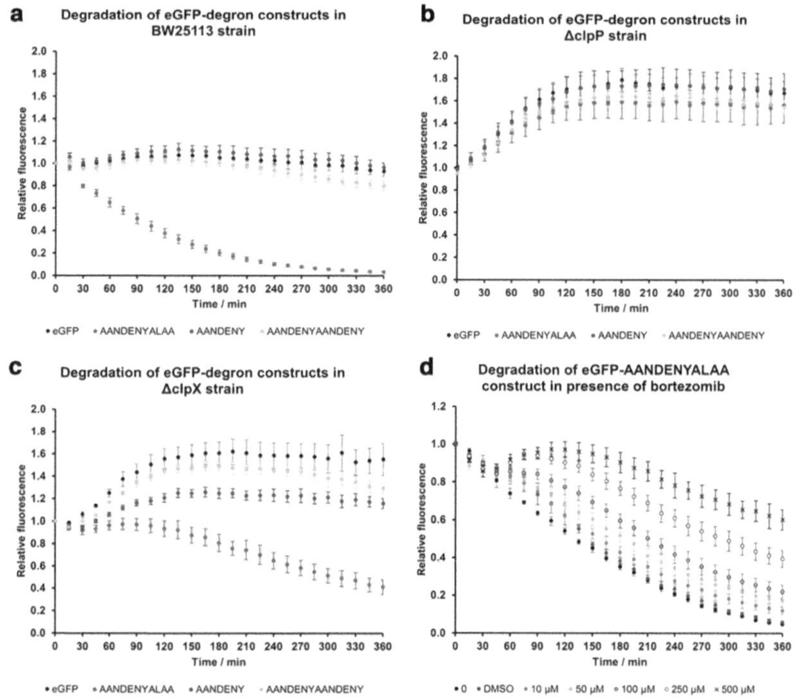

Fig. 3 Degradation of eGFP-degron constructs measured in liquid cultures in a 96-well plate format [6]. The plotted values represent mean of three biological replicates, and the error bars represent SEM. (**a**) Degradation of selected eGFP-degron constructs in wild-type BW25113 strain. (**b**) Degradation of eGFP-degron constructs in protease mutant strain ΔclpP. (**c**) Degradation of eGFP-degron constructs in unfoldase mutant strain ΔclpX. (**d**) Degradation of chosen eGFP-AANDENYALAA (ssrA) constructs in wild-type BW25113 strain in presence of increasing concentration of bortezomib

point. Then, normalize the data points to 100% at 0 min time point for each evaluated degron. Plot the processed data (normalized fluorescence vs. time) and determine $t_{1/2}$ of each eGFP-degron construct (*see* **Note 6**).

4 Notes

1. The protocol can be adapted to other plasmids, provided the chosen expression strain is compatible with the promoter (e.g., T7 promoter with *E. coli* strains DE3 expressing T7 RNA polymerase). Arabinose promoter of pBAD plasmid provides a tight expression control [8] and is compatible with expression in strains from the Keio collection [5], which provides an easy access to a wide variety of single-gene deletion mutants for validation of degron-specific proteases or adaptor proteins.
2. The described conditions were optimized for Phusion DNA polymerase. Other polymerases might require adjusting the temperature and time of the different PCR stages. If initial conditions do not yield enough of the DNA product or you observe a significant amount of non-specific product amplification, you might have to test other annealing temperatures to optimize the reaction conditions or pool several 50 μL reactions.
3. Depending on the purity of your DNA product, you might directly purify the DNA with a clean-up kit. If you observe the formation of non-specific DNA products, run the entire volume of the reaction on agarose gel (as above). Excise the band with the DNA product of expected length using a clean scalpel or a gel-cutting device and extract the DNA with a DNA gel-out kit.
4. An empty pBAD vector can be obtained by deletion of eGFP by site-directed mutagenesis using a protocol analogous to that described in Sect. 3.1. The primers for deletion of the eGFP should be complementary to the parts flanking the DNA sequence encoding the eGFP.
5. The kinetic interval and measurement time can be adjusted according to the degradation speed of specific degrons; however, evaporation of the sample will be more significant for longer plate incubations.
6. Plots can be analyzed using different data analysis software such as GraphPad Prism.

References

1. Varshavsky A (1991) Naming a targeting signal. Cell 64:13–15. https://doi.org/10.1016/0092-8674(91)90202-A
2. Bohn C, Binet E, Bouloc P (2002) Screening for stabilization of proteins with a trans-translation signature in *Escherichia coli* selects for inactivation of the ClpXP protease. Mol Genet Genomics 266:827–831. https://doi.org/10.1007/s00438-001-0601-1
3. Izert MA, Klimecka MM, Górna MW (2021) Applications of bacterial degrons and degraders—toward targeted protein degradation in bacteria. Front Mol Biosci 8:1–21. https://doi.org/10.3389/fmolb.2021.669762
4. Izert-Nowakowska MA, Klimecka MM, Antosiewicz A et al (2025) Targeted protein degradation in *Escherichia coli* using CLIPPERs. EMBO Rep 26:3994–4016. https://doi.org/10.1038/s44319-025-00510-9
5. Baba T, Ara T, Hasegawa M et al (2006) Construction of *Escherichia coli* K-12 in-frame, single-gene knockout mutants: the Keio collection. Mol Syst Biol 2:MSB4100050. https://doi.org/10.1038/msb4100050
6. Klimecka MM, Antosiewicz A, Izert MA et al (2021) A uniform benchmark for testing SsrA-derived degrons in the *Escherichia coli* ClpXP degradation pathway. Molecules 26:5936. https://doi.org/10.3390/molecules26195936
7. Schindelin J, Arganda-Carreras I, Frise E et al (2012) Fiji: an open-source platform for biological-image analysis. Nat Methods 9:676–682. https://doi.org/10.1038/nmeth.2019
8. Guzman L-M, Belin D, Carson MJ, Beckwith J (1995) Tight regulation, modulation, and high-level expression by vectors containing the arabinose P BAD promoter. J Bacteriol 177(14):4121–4130

Check for updates

Chapter 17

Assessment of Bacterial N-Degron Proteolysis by a Dual-Fluorescence Reporter

Pratistha Kandel, Patrick C. Beardslee and Karl R. Schmitz

Abstract

Many organisms possess N-degron pathways that link the proteolytic stability of individual proteins to the identity of their N-terminal amino acid. The set of N-degrons recognized by proteolytic machinery, and the regulatory paradigms leading to N-degron exposure, have been well studied in *Escherichia coli* and several other bacterial species. However, N-degron landscapes remain unexplored in the majority of bacterial phyla, and the extent to which these pathways vary among clades remains unknown. To characterize N-degron proteolysis in *Mycolicibacterium smegmatis*, a member of Actinomycetota and a model system for pathogenic mycobacteria, we devised a cell-based assay incorporating a dual-fluorescence reporter. We evaluated the proteolysis of mCherry substrates with defined N-terminal sequences by comparing fluorescence levels to a stable GFP reference in plate reader and flow cytometry contexts. By examining proteolysis patterns in wild-type *M. smegmatis* and cells lacking the N-recognin ClpS, we established that canonical primary destabilizing residues (Leu, Phe, Tyr, Trp) are subject to ClpS-dependent N-degron proteolysis. This dual-fluorescence approach is broadly applicable to interrogation of N-degron pathways across bacterial species.

Keywords N-degron, Proteolysis, Mycobacteria, N-end rule, Clp protease, Proteolytic reporter assay, Proteostasis

1 Introduction

Protein degradation is crucial for a wide range of physiological processes, ranging from protein homeostasis to regulation of individual cellular pathways. Proteolysis in the cytosol is performed primarily by ATP-dependent AAA+ proteolytic enzymes (ATPase associated with a variety of cellular activities). In eukaryotes, cytosolic proteolysis is a fundamentally centralized process carried out by a single protease, the 26S proteasome [1]. By contrast, bacteria possess multiple architecturally similar AAA+ proteases, which target separate sets of substrates and can be regulated individually. All ATP-dependent proteases possess unfoldase and peptidase activities, and share similar operating principles. For example, in Clp proteases (e.g., ClpAP and

Maria W. Górna et al. (eds.), *Degrons*, Methods in Molecular Biology, vol. 3069,
https://doi.org/10.1007/978-1-0716-5508-5_17,

ClpXP in *Escherichia coli*), one of the most widespread groups of ATP-dependent proteases in bacteria, substrate proteins are recognized by a Clp unfoldase (e.g., ClpA), mechanically unfolded using energy from ATP hydrolysis, and translocated into a barrel-shaped ClpP peptidase for hydrolysis [2, 3]. Untargeted proteolysis is inherently costly and potentially harmful, so ATP-dependent proteases have evolved the ability to selectively recognize substrate proteins from the complex background of the proteome. ATP-dependent proteases recognize some substrates through direct interactions with short peptide sequences, phosphorylated residues, or post-translationally appended degradation tags [4–10]. In other cases, substrates are delivered to the protease by accessory adaptor proteins [11, 12].

One well-characterized example of adaptor-mediated recognition occurs in N-degron pathways, where N-recognin adaptors recognize proteolytic substrates via specific N-terminal residues [13]. The set of residues recognized as N-degrons varies across organisms. In mammals, a large complement of normal and modified residues can serve as N-degrons, allowing multiple opportunities for proteolytic regulation of protein levels [13]. In the *E. coli* N-degron pathway, Tyr, Trp, Phe, and Leu comprise the primary destabilizing N-terminal residues [14]. This core set is expanded by amino-acyl transferases that prepend a destabilizing Leu or Phe in front of secondary destabilizing residues (Arg and Lys in *E. coli*; Arg, Lys, Asp, and Glu in *Vibrio vulnificus*) [15]. *E. coli* N-degrons are recognized by the ClpS N-recognin, which tethers substrates to the ClpAP protease for degradation [16–18]. Although non-essential, ClpS mediates rapid stimulation of N-degron degradation in vivo when present [19]. ClpS is also known to alter the substrate recognition by ClpAP by inhibiting the degradation of non-N-degron substrates [16]. The eukaryotic E3 ligase family shares a homology region with ClpS, suggesting that the mode of N-degron recognition is conserved in both prokaryotes and eukaryotes [20]. Similar N-degron pathways are known to occur throughout Proteobacteria and Cyanobacteria (as well as chloroplasts and apicoplasts), although the specific destabilizing residues recognized by ClpS are somewhat different among species, depending upon the selectivity of the ClpS binding pocket for hydrophobic side chains [21–28]. The existence of N-degron pathways in most other bacterial phyla has not been tested.

Interestingly, conserved ClpS orthologs are found in about half of Actinomycetota, a highly diverse bacterial phylum including the important human pathogen *Mycobacterium tuberculosis* [29]. The ~20 amino acid N-terminal segment of ClpS, which is required for transferring substrates to ClpA, is poorly conserved, whereas the globular ~75 residue core is well-conserved across Actinomycetota and shares ~30% sequence identity with *E. coli* ClpS [29]. A recent report demonstrated that *M. tuberculosis* ClpS interacts with the ClpC1 unfoldase and also delivers a model N-degron substrate to the essential ClpC1P1P2 protease for degradation in vitro [30]. Our recent study further confirmed that ClpS-dependent N-degron proteolysis occurs in the model mycobacterium *Mycolicibacterium smegmatis*, but only the four canonical primary destabilizing residues lead to degradation of model substrates [29].

Assessment of N-degron proteolysis in bacterial cells requires expression of model substrates with defined N-terminal amino acids. Complicating this is the fact that bacteria variably process protein N-termini. In *E. coli*, the formylmethionine used during initiation of protein translation is first deformylated. The resulting N-terminal Met is removed by methionine aminopeptidase if the second residue possesses a small side chain (Gly, Ala, Ser, etc.) but is left intact when the side chain of the second residue is bulky [31, 32]. This processing paradigm prevents constitutive exposure of N-degrons in nascent *E. coli* proteins. Moreover, the rules of leader Met processing are not well characterized in all bacterial groups, adding to the difficulty of controlling the identity of the leading residue. The foundational study that first probed the N-degron landscape in *E. coli* elegantly circumvented this uncertainty by fusing a ubiquitin domain in front of a β-galactosidase reporter [33]. Co-expression of a deubiquitinating enzyme resulted in removal of the ubiquitin shield and exposure of the following residue. This gave the authors a mechanism to control the N-terminal residue of the mature reporter.

To test for N-degron proteolysis in *M. smegmatis*, we constructed a dual-fluorescent reporter inspired by the ubiquitin-shielded constructs devised by Varshavski and colleagues. Our system utilized ^{HA}GFP-SUMO-X-mCherry^{myc} constructs driven by an anhydrotetracycline-inducible Ptet promoter in an integrating mycobacterial vector incorporating a ColE1 *E. coli* origin of replication, TetR cassette, L5 integrase, and KanR resistance marker derived from PLJR962 [34] (Fig. 1A). The mCherry reporter was decorated with a variable N-terminal tag with the sequence "X-LRVQGSTASGT," where "X" corresponds to one of the 20 amino acids and the remaining residues serve as a spacer that aids engagement and grip by proteolytic machinery [30, 35]. This tag was shielded by an upstream fusion with the *Saccharomyces cerevisiae* SUMO ortholog Smt3, analogous to the ubiquitin shield used in *E. coli* [14]. SUMO was chosen over ubiquitin because SUMO fusions are widely used as cleavable solubility-enhancing tags for recombinant protein purification [36], allowing us to use equivalent SUMO-X-mCherry constructs for in vitro studies. Constitutive cleavage of the SUMO shield was accomplished by co-expression of a downstream co-operonic gene encoding Ulp1, a SUMO protease from *Chaetomium thermophilum* [37, 38]. We also constructed a set of control constructs that lack Ulp1 (Fig. 1B), which do not expose the variable "X" residue.

To provide a co-translated non-proteolyzed fluorescent reference, an N-terminally HA-tagged superfolder GFP variant (GFP) was

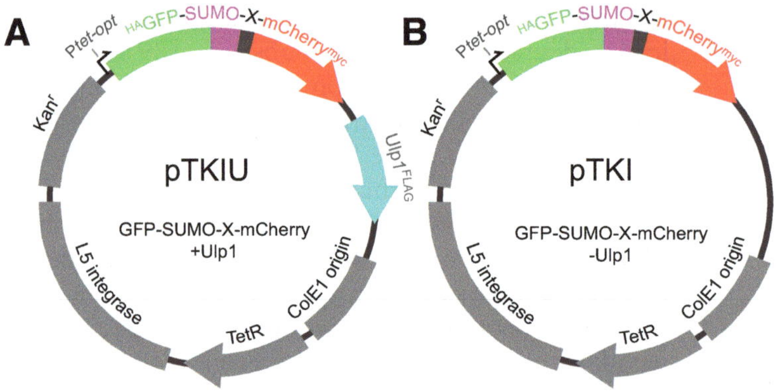

Fig. 1 Plasmid maps of reporter plasmids. (**A**) Map of the pTKIU plasmid. An anhydrotetracycline-inducible Ptet promoter drives expression of a bicistronic operon encoding the GFP-SUMO-X-mCherry reporter (where "X" is the N-terminal amino acid after removal of SUMO shield) and the Ulp1 SUMO protease. The ColE1 origin of replication allows episomal propagation in *E. coli*, while the L5 integration cassette directs genomic integration in the mycobacterial chromosome. (**B**) Map of the pTKI control plasmid, which lacks the Ulp1-encoding gene

fused to the N-terminus of SUMO [39]. Levels of the X-mCherry reporter were monitored by following the ratio of mCherry to GFP fluorescence, normalized by OD_{600} signal, rather than by following normalized mCherry signal directly. This ratiometric approach provided a more stable signal and made it possible to monitor changes in proteolysis across growth conditions where reporter expression may change (e.g., stress or starvation). Alternatively, the dual reporter allows two-channel analysis of cells by flow cytometry. Similar dual-fluorescent reporters have been used previously to monitor N-degron proteolysis in *E. coli* [40, 41]. The myc-tag on mCherry, HA-tag on GFP, and FLAG-tag on Ulp1 allowed us to confirm expression of all components by immunoblot.

To further evaluate the dependence of N-degron proteolysis on ClpS, we used CRISPR-Cas12a to generate a strain of *M. smegmatis* lacking the non-essential *clpS* gene [42, 43]. Similar to *E. coli*, deletion of *M. smegmatis clpS* did not affect cell growth under normal conditions [19]. In the following protocol, we use our dual-fluorescent reporter to compare steady-state proteolysis of model substrates presenting either N-terminal Tyr, a canonical primary destabilizing residue that leads to N-degron proteolysis, or Ser, which is not recognized as an N-degron. Similar approaches utilizing shielded dual-fluorescent reporters may be useful to investigate N-degron proteolytic landscapes in a wide variety of bacterial species.

2 Materials

2.1 Plasmid

1. pTKIU-Tyr: ^{HA}GFP-SUMO-Tyr-mCherry^{myc}; Ulp1^{FLAG} (Addgene ID: 247649) (*see* **Note 1**).
2. pTKI-Tyr: ^{HA}GFP-SUMO-Tyr-mCherry^{myc} (Addgene ID: 247647).
3. pTKIU-Ser: ^{HA}GFP-SUMO-Ser-mCherry^{myc}; Ulp1^{FLAG} (Addgene ID: 247650).
4. pTKI-Ser: ^{HA}GFP-SUMO-Ser-mCherry^{myc} (Addgene ID: 247648).

2.2 Strains

1. *Mycolicibacterium smegmatis* mc^{2}155 (wild-type; ATCC strain 700084).
2. Optional: *Mycolicibacterium smegmatis* mc^{2}155 (Δ*clpS*) (*see* **Note 2**).

2.3 Reagents

1. MB: Middlebrook 7H9 broth.
2. MBT: Middlebrook 7H9 broth, 0.05% w/v Tween-80.
3. MB agar: Middlebrook 7H9 agar.
4. Phosphate-buffered saline (PBS).
5. 50 mg/mL kanamycin (kan) stock solution in water.
6. 1 mg/mL anhydrotetracycline (aTc) stock solution in 50% ethanol.
7. Optional: purified mCherry and GFP.

2.4 Equipment

1. Plate reader with fluorescence and absorbance optics, temperature regulation, and plate-shaking capability.
2. Flow cytometer that supports dual-color fluorescence detection for GFP (e.g., 488 nm laser with 530/30 BP filter) and mCherry (e.g., 561 nm laser with 610/20 BP filter) with minimal spectral overlap.
3. Cuvette electroporator and electroporation cuvettes.
4. Black 96-well plate with clear bottom and clear microplate lid.
5. Sterile 14 mL culture tubes.
6. Sterile flow cytometry tubes.
7. Sterile 1.5 mL microcentrifuge tubes.

3 Methods

3.1 Transform Plasmids and Prepare Starter Cultures

1. Transform each reporter plasmid into *M. smegmatis* mc^2155 cells: Transfer 2 μL of miniprepped plasmid DNA into a prechilled 2 mm gap width electroporation cuvette. Add 50 μL of electrocompetent *M. smegmatis* mc^2155 cells. Transform by electroporation (2.5 kV). Promptly add 1 mL of MBT, transfer to a 14 mL culture tube, and incubate for 3 h at 37°C with 250 rpm shaking.
2. Transfer cultures to a sterile microfuge tube. Pellet cells at low speed in a microfuge and discard all but 50 μL of media. Gently resuspend cell pellets and plate on MB Agar + 20 μg/mL kan. Incubate plates for 3 days at 37°C, or until colonies form.
3. Prepare three (or more) replicate cultures for each sample from individual transformant colonies (*see* **Note 3**). Pick colonies and inoculate 5 mL starter cultures in MBT + 20 μg/mL kan in 14 mL culture tubes. Grow at 37°C with shaking for 3 days.
4. Subculture 500 μL of the starter culture into 5 mL of fresh MBT + 20 μg/mL kan and grow at 37°C with shaking until mid-log phase ($OD_{600} \approx 0.5$), ~16 h (*see* **Note 4**).

3.2 Assessment of N-Degron Proteolysis by Plate Reader Assay

1. Prepare 2× working stocks of cells: Replace media in each mid-log culture by centrifuging at 4000 × *g* for 5 min. Discard old media and resuspend in fresh MBT + 20 μg/mL kan at $OD_{600} = 0.1$.
2. Prepare a fresh 5 mL working stock of aTc at 100 ng/mL in MBT + 20 μg/mL kan (*see* **Note 5**).
3. Fill perimeter wells with water to control humidity: In a black 96-well plate with a clear bottom, fill wells along the outer perimeter with 200 μL of sterile water (*see* **Note 6**).
4. Prepare wells with induced samples: Add 75 μL of each 2× working sample stock to a well. Add 75 μL of aTc working stock to each induced sample well. Induced sample wells now contain cells at $OD_{600} = 0.05$ and 50 ng/μL aTc in a total volume of 150 μL.
5. Prepare wells with uninduced control samples: Add 75 μL of each 2× working cell stock to a well as before, then add 75 μL of MBT + 20 μg/mL kan to each. Uninduced sample wells now contain cells at $OD_{600} = 0.05$ in a total volume of 150 μL.
6. Prepare media-only wells: Add 150 μL of MBT + 20 μg/mL kan to media-only wells.
7. Optional fluorescent references: Prepare separate wells containing 50 nM purified mCherry or 100 nM purified GFP in media as standards for setting detector sensitivity (*see* **Note 7**).

8. Cover the plate with a transparent lid.
9. Transfer assay plate to a plate reader pre-incubated to 37°C. Monitor absorbance at 600 nm; GFP fluorescence at 468 nm excitation and 520 nm emission, and mCherry fluorescence at 587 nm excitation and 610 nm emission (*see* **Note 8**) at 10 min intervals over a duration of ~16 h at 37°C with periodic shaking (*see* **Note 9**).

3.3 Analysis of Plate Reader Data

1. Export raw data from the plate reader and import into a spreadsheet or data-processing software.
2. At each time point, subtract averaged absorbance and fluorescence values measured for the media-only wells from the respective absorbance and fluorescence values measured for sample wells.
3. For each sample, plot OD_{600} versus time to ensure that cells grow similarly across conditions and replicates (Fig. 2A). Large fluctuations or changes in growth rate at later time points may indicate changes in growth phase or cell clumping. If this occurs, data should be truncated to omit later time points.

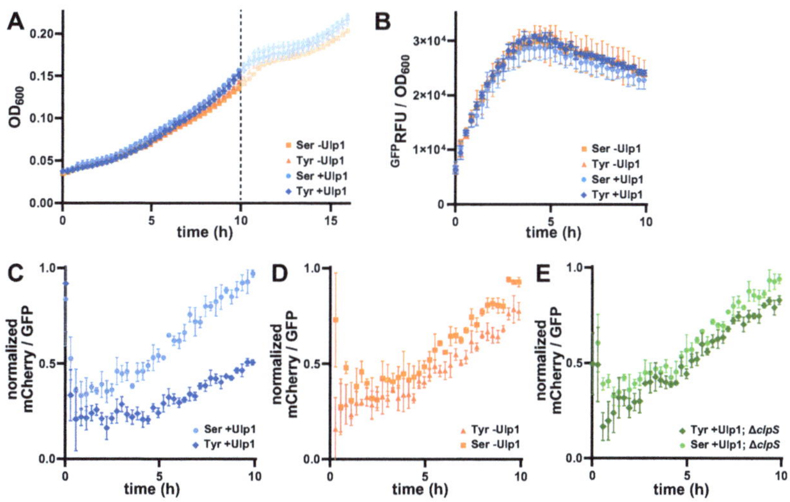

Fig. 2 Evaluation of N-degron proteolysis reporters in *M. smegmatis* by plate reader. (**A**) Growth curves of wild-type *M. smegmatis* expressing reporter constructs from pTKIU (Ser +Ulp1, Tyr +Ulp1) and control pTKI (Ser −Ulp1, Tyr −Ulp1) plasmids. All show similar growth patterns over 16 h at 37°C. Data were truncated to 10 h for further analysis, omitting the later phases where growth slows. (**B**) A plot of GFP fluorescence normalized to OD_{600} values shows comparable GFP fluorescence across samples. (**C**) Plotting the ratio of normalized mCherry fluorescence to GFP fluorescence shows lower steady-state mCherry levels for the reporter displaying an N-terminal Tyr, a canonical primary destabilizing residue, compared to Ser, which is not recognized by ClpS as an N-degron. (**D**) Cells expressing control plasmids in which the SUMO shield remains intact exhibit similar steady-state fluorescence ratios for Tyr and Ser constructs. (**E**) In cells lacking *clpS*, Tyr and Ser reporters accumulate to similar steady-state levels, illustrating the critical role of ClpS in N-degron proteolysis. Error bars indicate standard deviation of ≥3 replicates

4. At each time point, divide the media-corrected relative fluorescence units (RFU) by the respective media-corrected OD_{600} value, yielding $^{GFP}RFU/OD_{600}$ and $^{mCherry}RFU/OD_{600}$ values for each sample. RFU/OD_{600} traces can be plotted individually (Fig. 2B) (*see* **Note 10**).
5. Plot the mean and standard deviation of the ratio of $^{mCherry}RFU/OD_{600}$ to $^{GFP}RFU/OD_{600}$ values over time for each sample (Fig. 2C–E) (*see* **Note 11**).
6. Bar graphs of fluorescence ratios may be plotted for data from a single time point, or an averaged window of time points, from which unpaired *t*-tests can be performed for statistical analysis.
7. Expected outcome: The GFP-SUMO shield remains intact for control constructs expressed from pTKI plasmids lacking Ulp1, preventing recognition of N-degrons. Thus, for pTKI constructs, we correspondingly observe similar $^{mCherry}RFU/^{GFP}RFU$ traces irrespective of the identity of the "X" amino acid (Fig. 2D). By contrast, co-expression of Ulp1 from pTKIU plasmids results in removal of GFP-SUMO from the reporter, exposing the "X" amino acid. In the pTKIU context, RFU ratios reflect the accumulation of the Ser-mCherry reporter to higher steady-state levels than Tyr-mCherry (Fig. 2C), consistent with recognition and proteolysis of Tyr-mCherry by N-degron proteolytic machinery. In cells lacking *clpS*, RFU ratios are similar for both Ser and Tyr reporters expressed from pTKIU (Fig. 2E), validating the role of ClpS in recognizing Tyr and committing Tyr-mCherry to degradation.

3.4 Assessment of N-Degron Preference by Flow Cytometry

1. Inoculate liquid starter cultures of *M. smegmatis* harboring pTKIU GFP-SUMO-X-mCherry reporter constructs in MBT + 20 μg/mL kan. Grow with 250 rpm orbital shaking at 37°C until saturation (~3 days).
2. Subculture by 1:50 dilution into fresh media and grow at 37°C to mid-log phase ($OD_{600} \approx 0.5$).
3. Normalize subcultures to an OD_{600} of 0.05 in fresh media and induce reporter expression with 50 ng/mL aTc at 37°C in a shaking incubator for 10 h (*see* **Note 12**).
4. Centrifuge 3×10^8 cells (*see* **Note 13**) in sterile 1.5 mL microcentrifuge tubes for 5 min at $4000 \times g$. Discard supernatant and resuspend pellets in sterile PBS with 20 μg/mL kan. Store samples on ice until analysis.
5. Transfer cells from 1.5 mL tubes to flow cytometry tubes and load onto the cytometer. Set the event rate to 1000–2000 events/s. The event rate can be adjusted either by diluting the sample with more PBS or increasing the flow rate on the cytometer. Analyze 50,000 events once the flow rate is stable (*see* **Note 14**).

6. Monitor mCherry fluorescence by excitation with a yellow-green 561 nm laser (600 LP mirror: 610/20 BP filter) and GFP fluorescence by excitation with blue 488 nm laser (502 LP mirror, 530/30 BP filter) in flow cytometer (*see* **Note 15**).
7. In the flow cytometry software, generate a scatter plot with forward scatter area (FSC-A) on the x-axis and side scatter area (SSC-A) on the y-axis (Fig. 3A, B, left panels). Create a gate around the major clusters, excluding low SSC-A events (*see* **Note 16**).
8. Generate scatter plots of GFP fluorescence (y-axis) versus mCherry fluorescence (x-axis).
9. Expected outcomes: The scatter plot of fluorescence events generated by the N-degron Tyr-mCherry reporter exhibits similar GFP fluorescence values as the non-N-degron Ser-mCherry reporter, but mCherry fluorescence is clearly shifted to lower values (Fig. 3A, B, right panels).

4 Notes

1. Alternative fluorescent proteins may work as well or better than those chosen here. Fluorescent proteins may vary in folded stability and may be proteolyzed at different rates and efficiencies. Moreover, ideal reporter fluorophores must be spectrally distinct and should produce a strong signal at low expression levels, thus induction level and promoter strength may require optimization for ideal signal.
2. Comparative proteolysis of N-degron reporter substrates can be accomplished in wild-type cells. To test the role of ClpS in N-degron proteolysis, we additionally engineered a strain lacking *clpS* using CRISPR methods [42, 43].
3. It is advisable to make glycerol stock from each colony for long-term storage and use.
4. *M. smegmatis* behavior may change after long periods of time at saturation, and cultures should ideally be used during late log phase or soon after they reach saturation.
5. Anhydrotetracycline is light sensitive and is not stable in solution over the long term. For consistent induction behavior, solutions should be freshly prepared.
6. To mitigate evaporation of sample wells over multi-hour assays in a heated plate reader, we cover assay plates with a transparent lid and fill all outer wells with water. Consequently, only the 60 interior wells are available for samples and controls in a single plate.

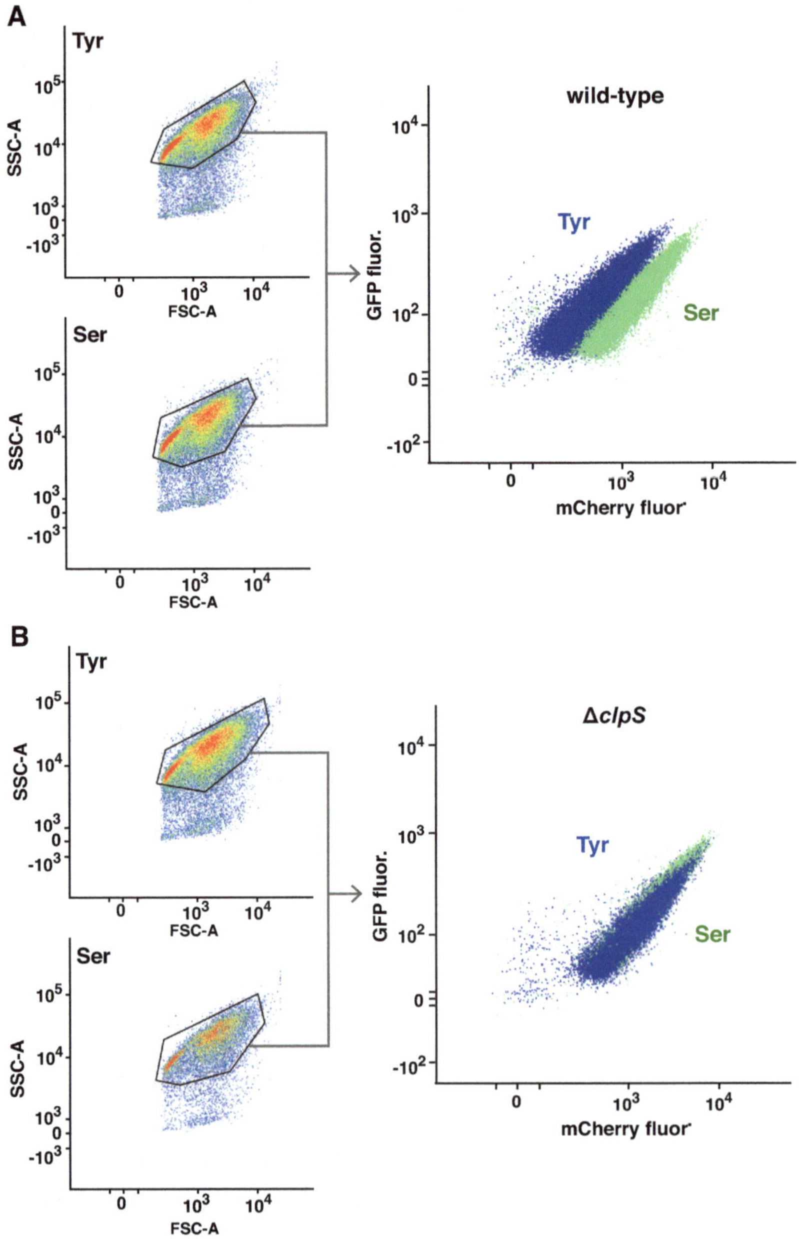

Fig. 3 Evaluation of N-degron proteolysis reporters in *M. smegmatis* by flow cytometry. (**A**) Wild-type *M. smegmatis* cells harboring either Ser or Tyr pTKIU plasmids were analyzed by flow cytometry after 10 h of aTc induction. Gates were established around the primary populations of cells in the SSC-A versus FSC-A plot (left), and the captured events were analyzed by plotting GFP versus mCherry fluorescence on a hyperlog plot (right). Cells expressing the Tyr reporter exhibit similar GFP fluorescence, but mCherry fluorescence is clearly shifted leftward to lower levels. (**B**) *M. smegmatis* Δ*clpS* cells expressing Ser or Tyr pTKIU plasmids were analyzed as in (**A**). GFP and mCherry fluorescence of both populations of cells overlap

7. Plate reader gain or sensitivity for fluorescence detection can typically be adjusted. For optimal results, these parameters should be optimized such that the maximum and minimum signals across the entire time course are within the dynamic range of the instrument. Alternatively, for instruments capable of calibrating gain against a reference well, we find it useful to include reference wells with purified GFP and mCherry. The concentration of purified protein appropriate to establish the ideal dynamic range must be determined for the individual assay conditions used.
8. Detection settings (wavelength, bandwidth, filters, etc.) for fluorescence detection may require optimization for the specific plate reader used.
9. Under these assay conditions, we find that incubation beyond 16 h is complicated by cell clumping and condensation on the plate lid, which cause unstable fluorescence and absorbance signals.
10. We observe that $^{GFP}RFU/OD_{600}$ values reach a maximum after ~5 h of induction, then gradually decrease as OD increases. Regardless of the specific trend observed, we expect the $^{GFP}RFU/OD_{600}$ values to be consistent across all samples.
11. At early time points, where cell concentrations are low and reporter expression has only just begun, there is greater uncertainty in both OD_{600} and RFU readings. Accordingly, RFU/OD ratios in the early portion of the experiments should be considered less reliable.
12. Induce for the optimal duration of time determined by the plate reader assay.
13. A culture of *M. smegmatis* at an OD_{600} of 1.0 corresponds to ~3×10^8 cells/mL.
14. Keep a stock of undiluted cells and prepare extra PBS in case fresh dilutions need to be made.
15. Our analysis utilized a nozzle size of 70 μm and neutral density (ND) filter of 1.0.
16. Bacterial populations may give multiple clusters on the SSC-A versus FSC-A plot, corresponding to different stages in the cell division cycle. With the *M. smegmatis* example data shown here, we found that these populations exhibited a consistent shift in fluorescence indicative of N-degron proteolysis and thus were gated as a single cluster.

References

1. Collins GA, Goldberg AL (2017) The logic of the 26S proteasome. Cell 169(5):792–806. https://doi.org/10.1016/j.cell.2017.04.023
2. Sauer RT, Baker TA (2011) AAA+ proteases: ATP-fueled machines of protein destruction. Annu Rev Biochem 80:587–612. https://doi.org/10.1146/annurev-biochem-060408-172623
3. Goldberg AL (1992) The mechanism and functions of ATP-dependent proteases in bacterial and animal cells. Eur J Biochem 203(1-2):9–23. https://doi.org/10.1111/j.1432-1033.1992.tb19822.x
4. Baker TA, Sauer RT (2006) ATP-dependent proteases of bacteria: recognition logic and operating principles. Trends Biochem Sci 31(12):647–653. https://doi.org/10.1016/j.tibs.2006.10.006
5. Mogk A, Schmidt R, Bukau B (2007) The N-end rule pathway for regulated proteolysis: prokaryotic and eukaryotic strategies. Trends Cell Biol 17(4):165–172. https://doi.org/10.1016/j.tcb.2007.02.001
6. Parsell DA, Silber KR, Sauer RT (1990) Carboxy-terminal determinants of intracellular protein degradation. Genes Dev 4(2):277–286. https://doi.org/10.1101/gad.4.2.277
7. Pickart CM (1997) Targeting of substrates to the 26S proteasome. FASEB J 11(13):1055–1066. https://doi.org/10.1096/fasebj.11.13.9367341
8. Pearce MJ, Mintseris J, Ferreyra J et al. (2008) Ubiquitin-like protein involved in the proteasome pathway of Mycobacterium tuberculosis. Science 322(5904):1104–1107. https://doi.org/10.1126/science.1163885
9. Burns KE, Liu WT, Boshoff HIM et al. (2009) Proteasomal protein degradation in mycobacteria Is dependent upon a prokaryotic ubiquitin-like protein. J Biol Chem 284(5):3069–3075. https://doi.org/10.1074/jbc.M808032200
10. Striebel F, Imkamp F, Özcelik D et al. (2014) Pupylation as a signal for proteasomal degradation in bacteria. Biochim Biophys Acta 1843(1):103–113. https://doi.org/10.1016/j.bbamcr.2013.03.022
11. Dougan DA, Mogk A, Zeth K et al. (2002) AAA+ proteins and substrate recognition, it all depends on their partner in crime. FEBS Lett 529(1):6–10. https://doi.org/10.1016/S0014-5793(02)03179-4
12. Sauer RT, Bolon DN, Burton BM et al. (2004) Sculpting the proteome with AAA+ proteases and disassembly machines. Cell 119(1):9–18. https://doi.org/10.1016/j.cell.2004.09.020
13. Varshavsky A (1996) The N-end rule: functions, mysteries, uses. Proc Natl Acad Sci U S A 93(22):12142–12149. https://doi.org/10.1073/pnas.93.22.12142
14. Tobias JW, Shrader TE, Rocap G et al. (1991) The N-end rule in bacteria. Science 254(5036):1374–1377. https://doi.org/10.1126/science.1962196
15. Balzi E, Choder M, Chen WN et al. (1990) Cloning and functional analysis of the arginyl-tRNA-protein transferase gene ATE1 of Saccharomyces cerevisiae. J Biol Chem 265(13):7464–7471. https://doi.org/10.1016/S0021-9258(19)39136-7
16. Dougan DA, Reid BG, Horwich AL et al. (2002) ClpS, a substrate modulator of the ClpAP machine. Mol Cell 9(3):673–683. https://doi.org/10.1016/S1097-2765(02)00485-9
17. Xia D, Esser L, Singh SK et al. (2004) Crystallographic investigation of peptide binding sites in the N-domain of the ClpA chaperone. J Struct Biol 146(1-2):166–179. https://doi.org/10.1016/j.jsb.2003.11.025
18. Zeth K, Ravelli RB, Paal K et al. (2002) Structural analysis of the adaptor protein ClpS in complex with the N-terminal domain of ClpA. Nat Struct Biol 9(12):906–911. https://doi.org/10.1038/nsb869
19. Wang KH, Sauer RT, Baker TA (2007) ClpS modulates but is not essential for bacterial N-end rule degradation. Genes Dev 21(4):403–408. https://doi.org/10.1101/gad.1511907
20. Lupas AN, Koretke KK (2003) Bioinformatic analysis of ClpS, a protein module involved in prokaryotic and eukaryotic protein degradation. J Struct Biol 141(1):77–83. https://doi.org/10.1016/S1047-8477(02)00582-8
21. Gao X, Yeom J, Groisman EA (2019) The expanded specificity and physiological role of a widespread N-degron recognin. Proc Natl Acad Sci U S A 116(37):18629–18637. https://doi.org/10.1073/pnas.1821060116
22. Román-Hernández G, Grant RA, Sauer RT et al. (2009) Molecular basis of substrate selection by the N-end rule adaptor protein ClpS. Proc Natl Acad Sci U S A 106(22):8888–8893. https://doi.org/10.1073/pnas.0903614106
23. Bouchnak I, Wijk KJv (2021) Structure, function, and substrates of Clp AAA+ protease systems in cyanobacteria, plastids, and apicoplasts: a comparative analysis. J Biol Chem 296:100338. https://doi.org/10.1016/j.jbc.2021.100338
24. Stanne TM, Pojidaeva E, Andersson FI et al. (2007) Distinctive types of ATP-dependent Clp proteases in cyanobacteria. J Biol Chem 282(19):14394–14402. https://doi.org/10.1074/jbc.M700275200
25. Nishimura K, Asakura Y, Friso G et al. (2013) ClpS1 is a conserved substrate selector for the chloroplast Clp protease system in Arabidopsis.

Plant Cell 25(6):2276–2301. https://doi.org/10.1105/tpc.113.112557
26. AhYoung AP, Koehl A, Vizcarra CL et al. (2016) Structure of a putative ClpS N-end rule adaptor protein from the malaria pathogen Plasmodium falciparum. Protein Sci 25(3):689–701. https://doi.org/10.1002/pro.2868
27. Tan JL, Ward L, Truscott KN et al. (2016) The N-end rule adaptor protein ClpS from Plasmodium falciparum exhibits broad substrate specificity. FEBS Lett 590(19):3397–3406. https://doi.org/10.1002/1873-3468.12382
28. Schuenemann VJ, Kralik SM, Albrecht R et al. (2009) Structural basis of N-end rule substrate recognition in *Escherichia coli* by the ClpAP adaptor protein ClpS. EMBO Rep 10(5):508–514. https://doi.org/10.1038/embor.2009.62
29. Presloid CJ, Jiang J, Kandel P et al. (2025) ClpS directs degradation of N-degron substrates with primary destabilizing residues in *Mycolicibacterium smegmatis*. Mol Microbiol 123(1):16–30. https://doi.org/10.1111/mmi.15334
30. Ziemski M, Leodolter J, Taylor G et al. (2021) Genome-wide interaction screen for Mycobacterium tuberculosis ClpCP protease reveals toxin–antitoxin systems as a major substrate class. FEBS J 288(1):111–126. https://doi.org/10.1111/febs.15335
31. Sherman F, Stewart JW, Tsunasawa S (1985) Methionine or not methionine at the beginning of a protein. BioEssays 3(1):27–31. https://doi.org/10.1002/bies.950030108
32. Ben-Bassat A, Bauer K, Chang SY et al. (1987) Processing of the initiation methionine from proteins: properties of the *Escherichia coli* methionine aminopeptidase and its gene structure. J Bacteriol 169(2):751–757. https://doi.org/10.1128/jb.169.2.751-757.1987
33. Hirel PH, Schmitter MJ, Dessen P et al. (1989) Extent of N-terminal methionine excision from *Escherichia coli* proteins is governed by the side-chain length of the penultimate amino acid. Proc Natl Acad Sci U S A 86(21):8247–8251. https://doi.org/10.1073/pnas.86.21.8247
34. Rock JM, Hopkins FF, Chavez A et al. (2017) Programmable transcriptional repression in mycobacteria using an orthogonal CRISPR interference platform. Nat Microbiol 2:16274. https://doi.org/10.1038/nmicrobiol.2016.274
35. Erbse A, Schmidt R, Bornemann T et al. (2006) ClpS is an essential component of the N-end rule pathway in *Escherichia coli*. Nature 439(7077):753–756. https://doi.org/10.1038/nature04412
36. Malakhov MP, Mattern MR, Malakhova OA et al. (2004) SUMO fusions and SUMO-specific protease for efficient expression and purification of proteins. J Struct Funct Genomics 5(1-2):75–86. https://doi.org/10.1023/B:JSFG.0000029237.70316.52
37. Lau Y-TK, Baytshtok V, Howard TA et al. (2018) Discovery and engineering of enhanced SUMO protease enzymes. J Biol Chem 293(34):13224–13233. https://doi.org/10.1074/jbc.RA118.004146
38. Mossessova E, Lima CD (2000) Ulp1-SUMO crystal structure and genetic analysis reveal conserved interactions and a regulatory element essential for cell growth in yeast. Mol Cell 5(5):865–876. https://doi.org/10.1016/S1097-2765(00)80326-3
39. Pédelacq J-D, Cabantous S, Tran T et al. (2005) Engineering and characterization of a superfolder green fluorescent protein. Nat Biotechnol 24(1):79–88. https://doi.org/10.1038/nbt1172
40. Kunjapur AM, Stork DA, Kuru E et al. (2018) Engineering posttranslational proofreading to discriminate nonstandard amino acids. Proc Natl Acad Sci U S A 115(3):619–624. https://doi.org/10.1073/pnas.1715137115
41. Sen S, Corrado N, Tiso AR et al (2025) Combinatorial mutagenesis of N-terminal sequences reveals unexpected and expanded stability determinants of the *Escherichia coli* N-degron pathway. bioRxiv. https://doi.org/10.1101/2025.05.22.655665
42. Yan MY, Yan HQ, Ren GX et al. (2017) CRISPR-Cas12a-assisted recombineering in bacteria. Appl Environ Microbiol 83(17):e00947–17. https://doi.org/10.1128/AEM.00947-17
43. Yan MY, Li SS, Ding XY et al. (2020) A CRISPR-assisted nonhomologous end-joining strategy for efficient genome editing in Mycobacterium tuberculosis. mBio 11(1):e02364-19. https://doi.org/10.1128/mbio.02364-19

Chapter 18

SsrA-Based Targeted Protein Degradation in *Escherichia coli*

Qichang Nie and Terrence Chi-Kong Lau

Abstract

Targeted protein degradation is an emerging concept of drug discovery to selectively eliminate the pathogenic proteins by activating their degradation in cells. The proteolysis-targeting chimeras (PROTACs) are bifunctional small molecules that induce the degradation of a protein of interest (POI) by proteasome. Here we describe a method that utilizes an ssrA-based BacPROTAC (bacterial PROTACs) to target the drug-resistant proteins CTX-M-14, a class A β-lactamase commonly found in extended-spectrum beta-lactamase (ESBL) plasmids of antimicrobial resistance (AMR) bacteria. This strategy could resensitize the drug-resistant bacteria and revive previously disregarded antibiotics, opens up a new avenue for therapeutic development of AMR bacteria.

Keywords BacPROTACs, Antimicrobial resistance, *Escherichia coli*, ESBL plasmids, Protein degradation, Proteasome

1 Introduction

In bacterial cells, ribosomal stalling occurs on mRNAs lacking stop codons or containing codons for which no cognate tRNA is available. To resolve stalled translation complexes and eliminate incomplete polypeptides, bacteria maintain a tmRNA-mediated quality control system (also referred to as SsrA RNA) in conjunction with its protein partner SmpB [1, 2]. This tmRNA (SsrA) system recognizes ribosomes arrested on truncated mRNAs and appends a degradation tag (SsrA tag) to the nascent polypeptide chain. The tag, featuring core sequences such as AANDENYALAA, is specifically recognized by ATP-dependent proteases such as ClpXP and ClpAP [3, 4], resulting in rapid proteolysis of the tagged substrate [5, 6]. This endogenous mechanism has been adapted as a tool for targeted protein degradation in *Escherichia coli*, enabling inducible depletion of specific proteins without genetic modification or exogenous chemical inducers in certain experimental systems.

Targeted protein degradation represents a pivotal methodology in molecular and cellular biology, enabling precise modulation of

Maria W. Górna et al. (eds.), *Degrons*, Methods in Molecular Biology, vol. 3069,
https://doi.org/10.1007/978-1-0716-5508-5_18,

protein abundance to investigate gene function, pathway dynamics, and therapeutic interventions. Current degradation strategies primarily engage one of two proteolytic systems: the ubiquitin-proteasome system or the lysosomal pathway. Chemically engineered molecules such as proteolysis-targeting chimeras (PROTACs) or molecular glues which bind specific target proteins to recruit the ubiquitin-proteasome machinery [7] or activate the lysosomal degradation pathway using lysosome-targeting chimeras (LYTACs), respectively.

In bacteria, the strategy of targeted protein degradation comprises two major approaches: degrader addition and degron fusion. BacPROTAC, the sole reported bacterial degrader, is a PROTAC-inspired molecule featuring a phosphorylated arginine analog (pArg) conjugated through a linker to a target-binding ligand (e.g., biotin). In bacteria, pArg binds the ClpC protease while biotin recruits target proteins, enabling proximity-mediated degradation via the ClpCP complex [8]. This approach has successfully achieved targeted protein degradation in *B. subtilis* and *Mycobacteria* and demonstrated therapeutic potential against bacterial infections and antibiotic tolerance [8–10]. Degrons, on the other hand, are specific peptide sequences that send the targeted proteins for proteasomal destruction. Gram-negative bacteria such as *Escherichia coli* often rely on endogenous proteolytic quality control mechanisms, and the sequence of SsrA tag (either at N- or C-terminus) has been demonstrated as the peptide sequence of degron with high efficiency, specificity, and versatility for targeted protein degradation [11, 12]. Moreover, the system of SsrA-mediated degradation provides several experimental advantages: it is genetically encodable, inducible via transcriptional control, and suitable for high-throughput applications [13, 14]. It has been employed to study essential genes, engineer synthetic genetic circuits, and interrogate protein function in real time. In this paper, we present a detailed protocol for SsrA-based targeted protein degradation in *E. coli*, including tag design, bacterial strain considerations, and validation methodologies.

2 Materials

2.1 Synthesis of Peptide Conjugates

1. Nacubactam: store at −80°C.
2. Dimethylformamide (DMF).
3. Trifluoroacetic acid (TFA).
4. HATU (1-[Bis(dimethylamino)methylene]-1H-1,2,3-triazolo[4,5-b]pyridinium 3-oxid hexafluorophosphate).
5. Piperidine (≥99.5%).
6. Dithiothreitol (DTT).
7. Fmoc-protected amino acids:

Fmoc-Ala-OH, Fmoc-Asn(Trt)-OH, Fmoc-Asp(OtBu)-OH, Fmoc-Tyr(But)-OH,Fmoc-Glu(OtBu)-OH,Fmoc-Leu-OH, Fmoc-6-Ahx-OH: store at room temperature.

8. Rink amide resin.
9. C-18 column.

All reagents are ACS-grade or synthesis-grade.

2.2 Cloning and Protein Expression

1. Antibody mouse monoclonal his-tag (sc-8036), HRP-conjugated m-IgG Fc BP-HRP (sc-525409), GAPDH monoclonal antibody (GA1R): store at 4°C.
2. *E. coli* BL21 (DE3) and J53 glycerol stocks: store at −80°C.
3. Plasmids: pACYCT2 and pET-28a.
4. Enzyme: NdeI and XhoI.
5. Antibiotics: 50 µg/mL kanamycin (Kan), and 34 µg/mL chloramphenicol (CHL).
6. Isopropyl β-D-1-thiogalactopyranoside (IPTG).
7. Ni-bound Sepharose resin.
8. Lysis buffer (pH 7.5): 150 mM NaCl, 20 mM Tris-HCl.
9. Wash buffer I (pH 7.5): 150 mM NaCl, 20 mM Tris-HCl, 20 mM imidazole.
10. Wash buffer II (pH 7.5): 500 mM NaCl, 20 mM Tris-HCl, 20 mM imidazole.
11. Elution buffer I (pH 7.5): 150 mM NaCl, 20 mM Tris-HCl, 200 mM imidazole.

3 Methods

3.1 SsrA-Based Conjugate Synthesis

1. Resin preparation: weigh 30.3 mg Rink amide resin (0.02 mmol equivalent) into a reaction vessel with filters and connect the up and down outlets to the solvent line and the waste line, respectively.
2. Before the first coupling step, add 8.0 ml DMF to the resin, stir gently for 5 min, and then drain the resin via nitrogen pressure (4–8 psi). Repeat three times.
3. Add 8.0 ml of 20% piperidine/DMF (v/v), stir gently for 5 min (The first Fmoc removal step), and then drain the resin by nitrogen pressure.
4. Dissolve 0.1 mmol Fmoc-protected amino acid and HATU (0.1 mmol), HOAt (0.1 mmol), and DIPEA (0.15 mmol) in DMF for 3 h HATU in 8.0 ml DMF. Mix well by bubbling nitrogen for 30 s.

Add the mixture to the resin. Shake for 1 h and then remove the solvent by nitrogen pressure (*see* **Note 1**).

5. Repeat the cycle starting from steps 3 and 4 for each of the subsequent amino acids according to the peptide sequence using the above Fmoc-amino acids.
6. Add different linkers before addition of nacubactam (*see* **Note 2**).
7. Peptide cleavage: Rink resin was stirred with cleavage solution (TFA/TIS/water = 95:2.5:2.5) at 0°C for 4 h.
8. Filter the mixture, and rinse and precipitate the crude peptide conjugates twice with cold diethyl ether, then centrifuge at 14,000 rpm at 4°C.
9. Peptide purification. The crude peptides were purified by RP-HPLC on a C-18 column at room temperature with a linear gradient system (5%–95% B in 6 min, flow 1 mL/min) using eluent A (100% H_2O + 0.05% TFA) and eluent B (100% CH_3CN + 0.05% TFA). The target fractions were collected, lyophilized, and confirmed by 4800 Plus MALDI TOF/TOF Analyzer (ABI).
10. Peptide powder were stored at −20°C or −80°C.

3.2 Cloning and Protein Expression

1. Transform the competent *E. coli* J53 cells with the pACYCT2 plasmid carrying the C-terminus his-tag ClpX sequence. Plate on 34 μg/mL chloramphenicol (CHL) agar plates and incubate overnight at 37°C.
2. Transform the competent *E. coli* BL21 cells with the pET28 plasmid carrying the C-terminus his-tag ClpP sequence. Plate on 50 μg/mL kanamycin (Kan) agar plates and incubate overnight at 37°C.
3. Target protein: β-lactamase, CTX-M-14, encoded by the *blaCTX-M-14* gene. The signal peptide coding region–truncated CTX-M-14 (sCTX-M-14) was inserted into pET-28a plasmid at NdeI and XhoI sites (*see* **Note 3**). Transform the competent *E. coli* BL21 cells with the pET28 plasmid carrying the signal peptide-truncated CTX-M-14 with N-terminus his-tag. Plate on 50 μg/mL kanamycin (Kan) agar plates and incubate overnight at 37°C.
4. Pick a single colony from the LB agar plate. All liquid cultures were grown under constant agitation (220 rpm) at 37°C.
5. Culture the cells on large scale and induce protein production. Incubate at 37°C with shaking until OD_{600} reaches 0.4–0.8. For most vector systems, induce with 500 μM IPTG and express protein for 3 h at 37°C, 5 h at 30°C, or overnight at 16°C (*see* **Note 4**).

6. Harvest the cells by centrifugation at 5000×*g* for 20 min at 4°C and discard the supernatant.
7. Collected cells are moved into a tube and flash-frozen in liquid nitrogen. For long-term storage, the sample is kept at −80°C or processed immediately (*see* **Note 5**).

All constructs are verified for their correct sequence by DNA sequencing.

3.3 Protein Purification

1. Resuspend the cell pellet into lysis buffer in a ratio of 1:10 (w/v) (*see* **Note 6**).
2. Lyse the resuspended cells using sonication (32% amplitude, 5 s on/5 s off, for 10 min).
3. Centrifuge the lysate at 12,000×*g* for 20 min at 4°C and discard the pellet.
4. Purify the protein through affinity chromatography (*see* **Note 7**). Equilibrate the Ni-NTA affinity column with 10 column volumes (CV) of lysis buffer.
5. Wash the resin with wash buffer I (5 CV).
6. Wash the resin with wash buffer II (5 CV).
7. Elute the recombinant protein with Elution buffer containing 100–200 mM imidazole.
8. Collect the samples and check the purity of the target protein in the fractions by SDS-PAGE.
9. Store the purified protein fractions (>95% pure, as estimated by SDS-PAGE) at 4°C for immediate use or frozen, at −80°C, for long-term storage.
10. Exchange the protein buffer with buffer (10 mM Tris-HCl, pH 7.5, 150 mM NaCl) after several centrifugation in 30K protein concentrator.
11. Protein concentrations were determined using the Bradford method.
12. Protein samples can be flash-frozen in liquid nitrogen and stored at temperatures below −80°C.

3.4 Binding Affinity with Protein Analyzed by Surface Plasmon Resonance (SPR)

1. The surface plasmon resonance experiments use a Bruker Sierra SPR-32 Pro Analyzer with a His-tag sensor chip.
2. Immobilize the ligands (sCTX-M-14, 1.5 μg/mL; ClpX, 4 μg/mL) in 10 mM Tris, 150 mM NaCl, pH 7.5 with 0.05% Tween 20 and 50 μM EDTA, separately onto the His-tag chip.
3. Block surfaces with a 7 min injection of 1 M ethanolamine, pH 8.0.
4. Activate the surfaces of flow cells for 7 min with a 1:1 mixture of 0.1 M NHS (N-hydroxysuccinimide) and 0.4 M EDC

(3-(N,N-dimethylamino) propyl-N-ethylcarbodiimide) at a flow rate of 5 μL/min.

5. To collect kinetic binding data, inject the analyte (NacssrA-1, nacubactam, or ssrA in 10 mM Tris, 150 mM NaCl, 0.05% Tween 20, and 50 μM EDTA, pH 7.5) over flow cells at concentrations of 90, 30, 10, 3.3, and 1.1 nM at a flow rate of 60 μL/min and 25°C.
6. The complex associates and dissociates for 90 and 300 s, respectively.
7. Regenerate the surfaces with a 5 s injection of 0.1 M EDTA. Triplicate injections (in random order) of each sample and a buffer blank flow over the surfaces.
8. Collect data at a rate of 1 Hz. The data fit to a simple 1:1 interaction model using the global data analysis option available within Bruker Sierra SPR-32 Pro Analyzer software.

3.5 In Vitro Protein Degradation Assays

1. Set up in vitro degradation assays: 0.5 μM *E. coli* ClpX, 0.5 μM *E. coli* ClpP, and 2 μM substrate protein, 15 μM phosphoenolpyruvate (PEP), 10 U/mL pyruvate kinase in a buffer (50 mM Tris pH 7.5 (at 37°C), 50 mM KCl, 20 mM $MgCl_2$, 10% (v/v) glycerol).
2. Dissolve BacPROTAC compounds in 100% (v/v) DMSO and further diluted, giving rise to a final concentration of 1% (v/v) of DMSO. DMSO as control.
3. Add 5 mM ATP to start the reaction and quench it by adding SDS sample buffer after 2 h of incubation at 37°C.
4. Analyze the degradation effect using SDS-PAGE and Coomassie blue staining. All experiments perform in triplicates.

3.6 Western Blot Analysis of Protein Degradation

1. Transform the competent *E. coli* BL21 cells with the respective pET28a vectors, culture the cells in LB medium (50 μg/mL kanamycin).
2. For in vivo degradation assays, induce protein expression by the addition of 20 μM IPTG.
3. Aliquot cells and incubate cells with continuous concentrations of NacssrA-1 or individual head molecules (nacubactam or SsrA peptide).
4. Harvest 200 μL of each treatment after 1-h incubation by centrifugation, resuspend the cell pellet in 100 μL of a buffer (50 mM Tris-HCl pH 7.5, 150 mM NaCl, and 10% (v/v) glycerol).
5. Lyse the cells using the Bioruptor (10 cycles, 30 s on / 30 s off) and centrifuge, get the cytosolic β-lactamase sCTX-M14 protein.

6. Analyze the samples using Western blot assay. Resolve 30 ng cell lysates of each sample by SDS-PAGE on a 10% gel at 100 V for 15 min and 120 V for 40 min. Transfer the protein blots to PVDF membrane (Sigma, German) at 100 V for 90 min at 4°C and block with 5% non-fat milk for 1 h at room temperature.
7. Incubate the membrane overnight with primary antibody: mouse monoclonal his-tag (1:1000), and GAPDH antibody (GA1R) (1:1000) at 4°C. Wash the membrane five times with TBST and incubate with secondary antibody (1:1000) for 1 h at room temperature, then wash again five times with TBST. Detect the signal on the membrane with an appropriate imaging system (e.g., ChemiDoc Touch Imaging System) and analyze using ImageJ.

3.7 Proteomic Analysis by LC–MS/MS

1. Sample preparation for LC–MS/MS: For quantitative MS analysis, treat *E. coli* BL21 (DE3) expressing sCTX-M-14 with 1% (v/v) DMSO, 10 μM, or 1 μM NacssrA-1, following the procedures described above.
2. Prepare the samples using EasyPep™ MS Sample Prep Kits (Invitrogen) following the manufacturer's protocol.
3. Measure the protein concentration by Rapid Gold BCA Protein Assay Kit (Pierce) using a CLARIOstar multimode microplate reader (BMG).
4. Take 15 μg of proteins from each sample, reduce and alkylate the protein at 95°C for 10 min using heat block.
5. After incubation, cool samples to room temperature, add trypsin/Lys-C proteases mix, and incubate for 3 h at 37°C.
6. Stop the digestion and transfer the digested protein samples to the protein peptide clean-up column for desalting and contaminants removal before drying using a Refrigerated CentriVap Vacuum Concentrator (Labconco).
7. Analyze peptide samples using a Bruker timsTOF Pro 2 mass spectrometer (Bruker, Bremen, Germany) interfaced with an UltiMate 3000 RSLCnano system (Thermo Fisher Scientific, Waltham, USA).
8. Reconstitute dried peptides in 30 μL of 0.1% formic acid in water and inject 2 μL of the solution onto an Acclaim™ PepMap™ 100 C18 trap column (5 mm × 1 mm; particle size, 5 μm; pore size, 100 Å).
9. Perform peptide separation on an Aurora™ ULTIMATE C18 analytical column (25 cm × 75 μm; particle size, 1.7 μm; pore size, 120 Å; IonOpticks, Fitzroy, Australia) using a 65-min linear gradient at a flow rate of 300 nL/min.
10. Initiate the gradient at 2% mobile phase B (acetonitrile with 0.1% formic acid) for 2 min, increase to 5% over 0.1 min, ramp from 5% to 35% over 44.9 min, then from 35% to 90%

in 4 min, and hold at 90% for 5 min. Subsequently, reduce the proportion of mobile phase B to 2% over 0.1 min and maintain for an additional 4.9 min.

11. Acquire the mass spectrometric data (including both MS and MS/MS spectra) over a mass range of 100–1700 m/z and scan the ion mobility range from 0.6 to 1.6 V·s/cm^2. The dual TIMS setup enables operation at 100% duty cycle, with ramp and accumulation times set at 100 ms. Collect MS/MS spectra in library-free DIA-PASEF (data-independent acquisition–parallel accumulation–serial fragmentation) mode, with a total cycle time of 1.8 s and an isolation window of 25 m/z. Collision energies range from 59 eV at 1.3 Vs cm^{-2} to 20 eV at 0.85 Vs cm^{-2}.
12. Analyze MS raw data using the Spectronaut™ 19 search engine (Biognosys Inc.) with the library-free (directDIA) workflow, utilizing a database comprising *E. coli* (strain B/BL21-DE3) and the sCXT-M-14.
13. Run the analysis with the default settings. In brief, perform the protein identification search using the following parameters set: (1) fixed modification: cysteine carbamidomethylation (+57.021 Da); (2) variable modification: methionine oxidation (+15.995 Da) and acetylation (+42.011 Da) at the protein N-terminus; (3) trypsin/P and Lys-C/P for digestion with two missed cleavages allowed; (4) Only PSM, peptides, and protein groups with false discovery rate (FDR) <1% were retained, and proteins with one or more unique peptides were filtered out.
14. Perform label-free quantification based on the identification search result with an FDR threshold of 1% at the MS2 level using the area of extracted chromatogram traces.
15. Visualize sCTX-M-14 protein within the *E. coli* proteome employing the pheatmap package in R. Analyze differential protein expression levels under the following criteria: a ratio >2 or <0.5 and a *P* value <0.05 with a focus on differential protein expression based on abundance ratios and significance thresholds.

3.8 Susceptibility Tests Using E. coli

1. Transform the competent *E. coli* BL21 cells with the respective pET28a vectors and culture the cells in LB medium (50 μg/mL kanamycin).
2. Harvest cells by centrifugation after 1 h of cultivation to obtain cellular protein levels suitable for monitoring within the dynamic range of detection of the Western blot experiment.
3. Perform susceptibility test of NacssrA-1 on sCTX-M-14–expressing bacteria by determination of the minimum

inhibitory concentration (MIC) of cefotaxime (CTX) using an agar-based assay.

4. Briefly, add 200 μL of molten LB agar supplemented with 50 μg/mL kanamycin and increasing cefotaxime (CTX) concentration at around 50°C to a 96-well microplate and cool down to room temperature.
5. Dilute cells in prewarmed LB medium to a final concentration of 1×10^5 CFU/mL, and spot 5 μL of cell dilution into each well and incubate at 37°C overnight before visual inspection.

4 Notes

1. When doing manual synthesis, intermittent stirring is preferred over continuous stirring because of the fragility of some resin, in order to minimize the fragmentation of the resin in small-sized particles, which can lead to the difficulties during filtration. As an alternative, continuous gentle shaking or nitrogen bubbling through the reaction vessel can be applied.
2. The length of the linker is a critical factor to modulate the binding affinity. In our study, three distinct spacers including a long 6-aminohexanoic acid (Ahx) linker, a short β-alanine linker, and no linker (control) were used to link the warhead (β-lactamase inhibitor, nacubactam) and an ssrA-degron peptide. Based on the structure prediction and molecular docking with the structure of CTX-M-14 (PDB: 6MZ1), ClpX (PDB: 3HTE), and the BacPROTACs using AlphaFold3, NacssrA-1 formed the most stable complex with a lowest binding energy. Therefore, Ahx, as a flexible linker, is the best option for bioconjugation of BacPROTACs.
3. Wild-type CTX-M-14 localizes to the periplasm of Gram-negative bacteria via an N-terminal signal peptide that directs translocation and is cleaved upon export to yield mature enzyme. In our study, we found that only the truncated CTX-M-14 (signal peptide–free protein) could fold correctly and form a soluble form. Thus, a signal peptide–free CTX-M-14 (sCTX-M-14) protein instead of the full-length protein is expressed.
4. Add 500 μM IPTG and express protein using optimal time/temperature determined in a small-scale trial.
5. Flash-freezing is recommendable before storing the cells but is not mandatory.
6. All the steps for cell lysis and solubilization should be carried out on ice, with all the buffers pre-chilled at 4°C.

7. The purification step should be done at low temperature (4–8°C) using a cold chamber or cold room. These steps can be performed using an FPLC system such as the ÄKTA purifier.

Acknowledgments

This work was supported by Research Grants Council, University Grants Committee (CityU 11102122), Health and Medical Research Fund, Food and Health Bureau, Hong Kong (22211142), Tung Biomedical Sciences Centre, City University of Hong Kong (9609322 and 9609308) to T.C.K.L.

References

1. Keiler KC, Waller PR, Sauer RT (1996) Role of a peptide tagging system in degradation of proteins synthesized from damaged messenger RNA. Science 271(5251): 990–993. https://doi.org/10.1126/science.271.5251.990
2. Karzai AW, Susskind MM, Sauer RT (1999) SmpB, a unique RNA-binding protein essential for the peptide-tagging activity of SsrA (tmRNA). EMBO J 18(13):3793–3799. https://doi.org/10.1093/emboj/18.13.3793
3. Gottesman S (2003) Proteolysis in bacterial regulatory circuits. Annu Rev Cell Dev Biol 19:565–587. https://doi.org/10.1146/annurev.cellbio.19.110701.153228
4. Gottesman S, Roche E, Zhou Y, Sauer RT (1998) The ClpXP and ClpAP proteases degrade proteins with carboxy-terminal peptide tails added by the SsrA-tagging system. Genes Dev 12(9):1338–1347. https://doi.org/10.1101/gad.12.9.1338
5. Klimecka MM, Antosiewicz A, Izert MA, Szybowska PE, Twardowski PK, Delaunay C, Gorna MW (2021) A uniform benchmark for testing SsrA-derived degrons in the Escherichia coli ClpXP degradation pathway. Molecules 26(19):5936. https://doi.org/10.3390/molecules26195936
6. Fei X, Bell TA, Barkow SR, Baker TA, Sauer RT (2020) Structural basis of ClpXP recognition and unfolding of ssrA-tagged substrates. Elife 9:e61496. https://doi.org/10.7554/eLife.61496
7. Békés M, Langley DR, Crews CM (2022) PROTAC targeted protein degraders: the past is prologue. Nat Rev Drug Discov 21(3):181–200. https://doi.org/10.1038/s41573-021-00371-6
8. Morreale FE, Kleine S, Leodolter J, Junker S, Hoi DM, Ovchinnikov S, Okun A, Kley J, Kurzbauer R, Junk L, Guha S, Podlesainski D, Kazmaier U, Boehmelt G, Weinstabl H, Rumpel K, Schmiedel VM, Hartl M, Haselbach D, Meinhart A, Kaiser M, Clausen T (2022) BacPROTACs mediate targeted protein degradation in bacteria. Cell 185(13):2338–2353. https://doi.org/10.1016/j.cell.2022.05.009
9. Won HI, Zinga S, Kandror O, Akopian T, Wolf ID, Schweber JTP, Schmid EW, Chao MC, Waldor M, Rubin EJ, Zhu JH (2024) Targeted protein degradation in mycobacteria uncovers antibacterial effects and potentiates antibiotic efficacy. Nat Commun 15(1):4065. https://doi.org/10.1038/s41467-024-48506-8
10. Junk L, Schmiedel VM, Guha S, Fischel K, Greb P, Vill K, Krisilia V, van Geelen L, Rumpel K, Kaur P, Krishnamurthy RV, Narayanan S, Shandil RK, Singh M, Kofink C, Mantoulidis A, Biber P, Gmaschitz G, Kazmaier U, Meinhart A, Leodolter J, Hoi D, Junker S, Morreale FE, Clausen T, Kalscheuer R, Weinstabl H, Boehmelt G (2024) Homo-BacPROTAC-induced degradation of ClpC1 as a strategy against drug-resistant mycobacteria. Nat Commun 15(1):2005. https://doi.org/10.1038/s41467-024-46218-7
11. Flynn JM, Levchenko I, Seidel M, Wickner SH, Sauer RT, Baker TA (2001) Overlapping recognition determinants within the ssrA degradation

tag allow modulation of proteolysis. Proc Natl Acad Sci U S A 98(19):10584–10589. https://doi.org/10.1073/pnas.191375298

12. Flynn JM, Neher SB, Kim YI, Sauer RT, Baker TA (2003) Proteomic discovery of cellular substrates of the ClpXP protease reveals five classes of ClpX-recognition signals. Mol Cell 11(3):671–683. https://doi.org/10.1016/S1097-2765(03)00060-1
13. McGinness KE, Baker TA, Sauer RT (2006) Engineering controllable protein degradation. Mol Cell 22(5):701–707. https://doi.org/10.1016/j.molcel.2006.04.027
14. Gur E, Sauer RT (2008) Evolution of the ssrA degradation tag in Mycoplasma: specificity switch to a different protease. Proc Natl Acad Sci U S A 105(42):16113–16118. https://doi.org/10.1073/pnas.0808802105

MIX
Papier aus verantwortungsvollen Quellen
Paper from responsible sources
FSC® C105338

If you have any concerns about our products,
you can contact us on
ProductSafety@springernature.com

In case Publisher is established outside the EU,
the EU authorized representative is:
Springer Nature Customer Service Center GmbH
Europaplatz 3, 69115 Heidelberg, Germany

Printed by Libri Plureos GmbH
in Hamburg, Germany